Stadia, Arenas and Grandstands

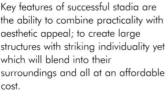

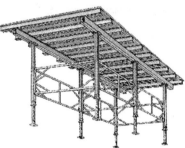

STADIA AND ARENAS

Development, design and management

Proceedings of the Second International Conference
'Stadia & Arena 2000'

Edited by J. J. A. Tolloczko and J. N. Clarke

THE STADIA AND ARENA INDUSTRY is faced by a period of challenges and opportunities – customers are expecting more exciting and sophisticated events, the IT revolution is having a tremendous impact on the management of stadia and arenas, and on broadcasting media, and architects and engineers are bringing tremendous imagination and creativity to the design and construction of new stadia throughout the world.

The book forms the proceedings of the Second International Conference *'Stadia & Arena 2000'* held at the London Arena in June 1999. It provides all those involved in the development, design and management of new and existing stadia and arenas with a wide-ranging view of the challenges and opportunities ahead.

This book brings together more than 30 expert overviews and case studies of key topics in this growing field, grouped under seven themes:

◆ Planning and safety

◆ Development, operation and finance

◆ Design and construction

◆ Grass and playing surfaces

◆ IT and services

◆ Managing venues

◆ Catering and hospitality.

The Concrete Society, June 1999
Paperback, 304 pages, 234×156 mm
ISBN 0 946691 73 8
Price £55

To order, please contact:
The Concrete Society, Century House, Telford Avenue, Crowthorne, Berkshire RG45 6YS, UK
Tel: +44 (0)1344 466007, Fax: +44 (0)1344 466008
Email: concsoc.concrete.org.uk, http://www.concrete.org.uk

SPONSORED BY

IN ASSOCIATION WITH

ORGANISED BY

Stadia, Arenas and Grandstands
Design, construction and operation

Proceedings of the First International Conference
"Stadia 2000"

Cardiff International Arena, Cardiff, Wales, 1–3 April 1998
Organised by The Concrete Society

EDITED BY

P. D. Thompson
Millennium Stadium plc

J. J. A. Tolloczko and J. N. Clarke
The Concrete Society

London and New York

This edition published 1998
by E & FN Spon, an imprint of Routledge
11 New Fetter Lane, London EC4P 4EE

Simultaneously published in the USA and Canada
by Routledge
29 West 35th Street, New York, NY 10001

Reprinted 2000

Spon Press is an imprint of the Taylor & Francis Group

Printed and bound in Great Britain by
St Edmundsbury Press Ltd, Bury St Edmunds, Suffolk

Publisher's Note
This book has been prepared from camera-ready copy and
word-processing discs provided by the individual contributors.

British Library Cataloguing in Publication Data
A catalogue record for this book is available
from the British Library

ISBN 0 419 24040 3

CONTENTS

ADDITIONAL PAPERS

Additional papers that were presented at the Conference in Cardiff from 1 to 3 April 1998 are listed below. These were not available for inclusion in this present volume but copies are available from The Concrete Society. For more information, contact: The Concrete Society, Century House, Telford Avenue, Crowthorne, Berkshire RG45 6YS, UK. Tel: +44(0)1344-466007, Fax: +44(0)1344-466008, Email: concsoc@concrete.org.uk.

Key Note Address: Rugby Union: From a players game to a sporting spectacular
While Rugby at the grass-root level remains an amateur pastime played for enjoyment by millions, at the top professional level it is turning into a spectators sport for the enthusiastic follower. This dramatic change has considerable implications for the traditional Rugby stadium. This new approach to the game is reflected in the emergence of astounding high-tech sporting venues such as the Millennium Stadium Cardiff - arguably the most advanced Rugby venue in the world.
V. Pugh QC, Chairman, International Rugby Board

Stadia: the sleeping giants of tourism
The presentation will discuss the potential for tourism to benefit from new stadia developments and will explore the interactions between the two activities which have not operated as effectively together as is desirable.
Dr T. R. Stevens, Stevens Associates, Swansea, UK

Managing stadia: the human resource issues
The paper will outline a comprehensive human resource strategy developed for those employed in the management and development of the stadia and arena sector of the leisure industry.
Dr T. R. Stevens, Stevens Associates, Swansea, UK

Revenue and access control systems for the stadia and arena industry
The presentation will discuss customer challenges in relation to ticketing, and outline the wide range of new technology solutions now available for all types of event at stadia and arenas.
W. R. Stapleton, Lasergate Systems Inc., Clearwater, Florida, USA

European standardisation work on spectator facilities
Standards for architectural design and performance requirements for spectator facilities for sports and multipurpose venues are in preparation to ensure safety, comfort of, and visibility for spectators. Permanent indoor venues (theatres, cinemas, etc) are excluded. Coverage will include: standards for layout criteria including spacing, access and egress, sight lines, positioning of separation fences and barriers; and standards for products by performance requirements for permanent, demountable, movable telescopic stands.
J. Poyner, Chairman, BSI Committee B/522, and Vice-President, Hussey Seating Co., Warrington, UK

Design and build: the integrated approach

Design and build is a forceful contender among the numerous choices of contract available to clients when procuring their project. Project success is underwritten by having an integrated team who all understand the value in the completed project serving the public on the appointed date, thus guaranteeing the client' targets. A range of sports stadium developments will be illustrated from design brief, through construction to operating the facilities.

W. J. Barr and A. G. Geddes, Barr Construction, Birmingham, UK

PREFACE

The market for specialist and multi-functional sports and leisure facilities to host international, national and regional sports, conferences and leisure events is growing fast. This book presents the best of current experience in the design, construction and operation of these exciting structures. It includes contributions from experts in many fields of procurement, planning and design, and case studies of lessons and opportunities.

The book forms the proceedings of the First International Conference & Exhibition on the *Design, Construction & Operation of Stadia, Arenas, Grandstands & Supporting Facilities* held in Cardiff on 1–3 April 1998 under the title *"Stadia 2000"*, and organised by The Concrete Society. This major event captures the excitement of the growing market for specialist and multi-functional sports and leisure facilities. In particular it brings together all concerned to exchange views and establish contacts and business opportunities.

Among projects covered in detail are stadia and arenas in Europe, North America, Japan and the Far East. Among the individual stadia which are described are some of the most evocative names in sport: Cardiff Arms Park, Wembley, Wimbledon, Twickenham. But there is no doubt that many of the new generation of stadia, which are described by their designers and constructors in the chapters which follow, will become as well-known in years to come.

Emphasis is of course now rightly placed on safety and on providing spectators and customers with an enjoyable experience. Environmental and aesthetic issues, and the high-tech approach to stadium management and facilities are of increasing importance, especially in the competitive commercial world.

Thanks are due to the sponsors of the event for their support: Bison, Philips Projects, British Steel, Millennium Stadium, Rugby World Cup 1999 and *Stadium & Arena Management*, and to the Conference Committee for their advice and assistance.

Pat Thompson
Jurek Tolloczko
Nick Clarke

February 1998

CONFERENCE STEERING COMMITTEE

FOREWORD

STADIA AND ARENA THROUGH THE AGES

R. K. SHEARD
Lobb Sports Architecture, London, UK

INTRODUCTION

I am a venue lover, addicted to the design of venues like others are addicted to drugs.

I suspect some of you may be also. You will recognise the symptoms – you go to an event and you spend more time watching the crowd than the players, you watch how the stairs work and how the crowd flows, what people are buying at the concession and how long they are waiting. And what are you thinking all the time – how can it be done better?

Well that is why we are here for this conference – to learn from others: we all have knowledge which will be useful to our neighbours, so now it is my turn – I hope it is useful.

Definitions

I will use the word stadia quite often but I use it to describe all sports venues whether they are outdoor open stadia, indoor arenas or glass-enclosed racecourse grandstands. Wherever the public come to be 'entertained' by sport. I say sport because although it is not the only function of these buildings, it is the common bond and usually the main function of most of these venues.

Stadia

They can be wonderful places, they can host the population of a city for a few hours and can contain all of the good and bad in that part of society who are attracted to physical skill and endurance, and more recently vocal effort. They can be spectacular, uplifting buildings that punctuate our lives with enjoyment and sometimes disappointment. They can be places of worship and inner thought and they can bind us with our fellow man like no other building.

On the other hand they can be dark, inhospitable places, unsafe and insecure. A disappointment to our expectations.

Expectations

But what are our expectations of these venues? The ideal stadium is clearly the one in which our team always wins. But what do people really want from these buildings, what do people expect from these wonderful venues, these 'cities for a day'? I believe they want comfort and expect safety. A building which simply allows us to be comfortable in our enjoyment of the event, a building which is safe for us to become part of the crowd, to feel comradeship, togetherness, part of something big. These two factors, comfort and safety, are part of the success formula for any stadia. All three of these factors must be present for a successful development

Involvement
For us to design comfortable and safe buildings we need to understand the spectators and the emotional cycle they go through before, during and after the event: how this cycle blends with the cycle of the event itself. Involvement is what people go to see sport for but once they are involved, they are part of a crowd with the potential for good or bad, just like the rest of society.

Facility complex
So how do we achieve these ideals? – we plan, test, have ideas, develop and think through the solutions, what we don't do is just copy the last thing we saw in a magazine.

Just look how the complexity of stadia have changed. Before we can gaze into the future we have to know where we have come from and where we are now.

THE DEVELOPMENT OF STADIA

First generation
I believe there have been three generations of stadia design. The first was brought about by the 'rules' revolution in the mid 19th century when the informal games which were then played by small communities and the landed gentry were organised and the rules of modern sports were formed. This organisation of sport led to a peak of spectators watching live sport in the mid-twentieth century.

It was the first generation of the modern stadium, where ticketing was the key to financial success; 'bums on seats' was the catch-phrase.

Second generation
Then in 1937 began the 'television' revolution which broadcast live sport around the world to an increasing armchair audience, while audiences at live sport started to decrease. To compete with this threat, the second generation of stadia evolved which recognised that to attract that live audience back to the stadia they had to be given more comfort and more safety. The second generation of stadia placed a higher priority on concourses which were used to full advantage. Spend per head was the key.

Third generation
Today we are at the beginning of the third revolution in sport, the 'entertainment' revolution, when sports are having to accept that they are now big business, competing for our leisure time along with the many other forms of entertainment. This is producing the third generation of stadia where service is the key.

The third generation is not designed for summer or winter, for football or rugby, for young or old; they are designed for as wide a range of events and audience as possible. Entertainment is the aim and 'service' is the method.

THE FUTURE

So what does the future hold ? What will be the fourth generation? The next generation of stadia, I believe, will evolve from advances in technology. Technology is allowing us as designers to smooth out the differences between the sporting codes in their use of pitch surfaces and movable stands. It also allows us to minimise the difference between sports and other events by the use of movable pitches and moveable roofs.

Pitch

The line between natural grass and synthetic pitches is merging with developments in plastic mesh root reinforcement and plastic turf support with computer-controlled nutrient injection. Combined with this, new hybrid grass types require less light, grow faster and are far more robust.

Seating

Seats are changing, now being ergonomically designed with integral padding of soft plastics bonded in manufacture. These will give the effect of a padded seat but in a seamless and weatherproof shell. In one arm will be the sockets for plugging in personal headsets to listen to 'Stadium Radio' or 'Stadium Television' if the receiver is hired. As the average population size increases, the seat spacing will also increase and allow for slight adjustment of seat backs to suit the individual. Pockets on the back of the seat in front will contain the free 'Stadium Catalogue', advertising products on sale by post or from the Stadium Retail Centre. Items will be able to be ordered using the hand-held receiver hired for the day, purchases can be waiting for you at the shop at the end of the match or delivered to your seat at half-time.

Viewing standards

Viewing standards defined by sight lines for most stadia are now calculated on computer and the creation of three-dimensional computer models of developments will allow spectators to see exactly the view they will have from their seat at the time of booking.

Facilities

Support facilities will provide amenities for all the family to enjoy, as well as other entertainment areas for those not committed to the event. They will eventually include every type of function from business centres to video game arcades, similar to the range of facilities found in an international airports. Attractions will be designed to encourage spectators to arrive early and stay on afterwards, perhaps even sleeping overnight in the Stadium Hotel.

Tomorrow's stadia will be places of entertainment for the family, where sport is the focus but not the complete picture. It will be possible for five members of a family to arrive and leave together but in the intervening period experience five different activities. While the parents 'see' the live game their children 'experience' the live game in the virtual reality studio where images from the 'in pitch' cameras provide close, immediate action.

Information

Information technology will provide the spectator with all the advantages of the television viewer at home. Information is needed to keep the audience knowledgeable about the event. This knowledge entertains and extends the attention span which prevents boredom in the crowd. It can be used to attract spectators into the stadium early and keep them later, reducing the pressure on the circulation system. This peripheral information is particularly important when we recognise that the period of actual play is often only a fraction of the full event.

One set of tennis at Wimbledon, for example, which lasts 30 minutes, may only have four or five minutes of actual play. A five-set match which lasts two and a half hours may only have 20 to 25 minutes of play. A football match which lasts one and a half hours may again only have 20 minutes of action, leaving plenty of time for the attention to wander.

Communication

The aim of the fourth generation stadium will be to offer information equal to professional broadcasts at facilities which are as safe and comfortable as our own homes. Replays, information about players, previous match highlights, statistics on the game, expert commentary and even advertising will be stored on the stadium database. This is possible through 'narrow casting' by the stadium's own CCTV network, eventually by satellite, not just to one or two large video screens but to small personal receivers with screens a few inches across. These receivers will be part of the ticket price or on hire for the day and will only receive the stadium channel; probably with ear phones and eventually with interactive controls allowing a choice of information. Press the 'statistics' button and type in your favourite player's name and his career statistics will be displayed, press 'action' and type in the date of the match and see the highlights of his match-winning performance two years ago.

Active pass

With the introduction of the 'active pass', conventional turnstiles will disappear and venues will benefit from the 'intelligent' entrance gate linked to the stadium computer system. From these access points, which will look more like the x-ray machines used at airports, full details will be read from the spectator's active pass. The pass will be automatically scanned by monitors and if the pass is invalid a warning will sound and the holder advised by synthesized voice where to go to seek help. The automatic crowd control barriers in front, which are usually open, will close if the person attempts to proceed any further. Each pass will allow the spectator access to different areas of the ground and entitle the holder to other pre-determined benefits.

In addition to this access and sales control, the computer will store information such as age, sex, address and event preferences of the spectators. From this database of information, management will be able to form an exact profile of who attends what events, which will allow them to target that exact socio-economic group the next time a similar event is held. This knowledge of the spectator is essential for future marketing.

Television

The last and most important change technology will bring about is digital television, which I believe will have a profound affect on the fourth generation stadia. This technology will allow TV companies to broadcast not just three or four channels from a satellite, but three or four hundred. The total number of television hours will explode around the world with companies needing to fill their 'on air' hours. Sport is the natural answer. It is cheap to produce and almost always attracts an audience.

Sport is cheap television programming for the digital age. Stadia will evolve which uses this boom in television hours where the financial viability of the venue will not be dependant upon those through the gates or even how much they spend, but those sitting at home.

The fourth generation will be where the 'Stadium Studio' is the answer and technology is the key.

CONCLUSION

It wasn't until the 1932 Los Angeles Olympic Games that technology was used to determine the result of an event for the first time and the Kirby Photo-finish Camera took an hour to produce the result. Technology is now essential to the smooth operation of a venue. We expect races to be timed to thousandths of a second, drug samples to be analysed in laboratories to particles per million and video play-backs provided instantly. But this is only the tip of the technological ice-berg. As an architect, I am confident about finding solutions to design problems society has set us by applying new technology. There are problems, however, which are part of the fabric of society itself, which cannot be solved by design. When the Arsenal North Bank Stand opened in 1993, it was described as social engineering, but this is not the case. Our solutions for Arsenal were unusual, some may say innovative, but they are solutions society has been requesting for years.

Stadia design has not kept pace with the changing needs of a sophisticated society; it is time we listened to what people really want from our venues, and give it to them.

PART ONE

DEVELOPMENT AND PLANNING

1 STADIUM PLANNING IN TODAY'S TOWNSCAPES

P. ROBERTS
Fuller Peiser, UK
M. DICKSON
Buro Happold, Bath, UK

SUMMARY: This paper examines the impact of the Taylor Report and recent Government legislation in the UK, the social, planning and design complexities of stadia design and, through recent examples and discussion of current design technology, suggests a way forward for the creation of imaginative and sustainable stadia for the 21st century.

Keywords: All seater stadia, brand opportunity, evacuation in fire, innovation, land reclamation, local community, multi-functioning, planning authority, policy review, siting constraints, Taylor Report, townscape.

INTRODUCTION

Building or refurbishing a stadium is an opportunity to create a landmark structure of technical excellence, which can serve to enrich the local community and provide economic benefits and prestige to the owning club and the local area.

Sadly, much recent stadia development in the UK has fallen short of that developed by other cultures. This is despite Lord Justice Taylor's wide-ranging report of 1990 which contained "recommendations about the need for crowd control and safety in sports events" [1]. Taylor's report stressed the need for such facilities to be flexible, providing a 'social service' with wide usage within the community.

This paper examines the impact of the Taylor Report and recent Government legislation in the UK, the social, planning and design complexities of stadia design and, through recent examples and discussion of current design technology, suggests a way forward for the creation of imaginative and sustainable stadia for the 21st century.

PLANNING ISSUES

Recent policy and guidelines

Stadia are complex entities. With a huge turnover of visitors and a strong visual presence, a stadium development also affects the surrounding area in terms of traffic, economic and environmental impact.

Throughout the 1960s and 1970s, the issues surrounding planning of stadia were not

Stadia, Arenas and Grandstands, edited by P.D. Thompson, J.J.A. Tolloczko and J.N. Clarke.
Published in 1998 by E & FN Spon, 11 New Fetter Lane, London EC4P 4EE, UK.   ISBN: 0 419 24040 3

fully examined; no separate policy existed for design of stadia. In the late 1980s, however, the Government started to produce planning policy guidance notes (PPGs) on specific topics, the most memorable being on green belts and housing. In the early 1990s, guidance on Sport and Recreation (PPG17) finally appeared, and this happened to coincide with the outcome from the Taylor Report of 1989.

The Taylor Report marked the first major examination of the issues surrounding siting and design of stadia and led to a surge of activity in sports related development. The issues discussed as a result of the Hillsborough Disaster led to the following Government guidelines:

- the necessity for close liaison at an early stage between local authorities and sports clubs wishing to develop stadia
- the need to view the siting of a major new stadium as an important strategic issue in the overall development plan
- proper consideration of access to public transport, parking and possible conflicts with neighbouring users when re-developing and improving existing stadia
- importance of multiple use, sustainability and community benefit
- encouraging the use of reclaimed and derelict land for siting when a particular club chooses to relocate.

However, during a stage of heavy investment in sporting facilities and in stadia development, it was discovered that this policy approach did not take proper account of the complexity of stadia development. Resulting schemes tended to be large, were often promoted on greenfield sites, caused transport concerns, did not consider the significant visual impact and, due to lack of long-term planning, were often developed in an atmosphere of poor relations with both the local community and authority.

The local community: planning process issues

It is vital for a new stadia development to exploit the opportunities of good relations with the local community and authority. Sports clubs have a great cultural bond with the local area and this can help to win support. The visual, social and economic benefits of working to enrich the surrounding area will be examined in the second half of this paper.

The local authority in Huddersfield, Kirklees, promoted the scheme for a local stadium, underwrote the development and took a share in the stadium company, in an example of wholehearted backing by a local authority at its most constructive. In another instance, Sunderland City Council used its unitary development plan to identify a new site when it became clear that Roker Park needed development.

The necessary close ties with the local authority can also cause problems, where over-zealous authorities refuse planning permission or restrict design opportunities. At Airdrieonians Football Club, the club identified 13 sites between 1992 and 1994 and, with Lanarkshire Development Agency and Monklands District Council, undertook extensive feasibility studies. Despite this, all sites were rejected and a further four were submitted and turned down before the Planning Authority finally granted permission. Even at this stage, the scheme had gathered 220 objections and a residents' group had been set up to oppose the scheme. Where local feeling runs high, similar problems can occur: in Edinburgh, a City study failed to locate a suitable site for the controversial amalgamation of Hearts and Hibernian football teams.

Siting and the need for a development plan

Right from the start, it is necessary for local authorities and clubs to work in partnership, identifying both if a new site is necessary and then where it should be, while working with the statutory planning process at every stage. Where site specific issues are concerned, any scheme can be thrown out if its siting does not comply with policy.

Development plan policy, or lack of it, can subject schemes to call-in inquiries. Dewsbury Rugby League Club is a classic example. Whilst it was in the Green Belt, the scheme was supported by the Local Authority and, notwithstanding a lack of specific policy guidance for this type of development, only attracted three letters of objection. However the application was called-in for determination by the Secretary of State for the Environment the day before the planning committee, despite a recommendation for approval. Subsequently, a Public Inquiry occurred where the Inspector recommended refusal. The Secretary of State disagreed with the Inspector's decision after satisfying himself that no other options existed. This clearly demonstrates the problem of a policy vacuum.

The aim for a successful lasting facility is often compromised by the following:

- stadia are very complex in terms of structure and internal planning
- clubs often have no experience or knowledge of planning and urban issues
- on occasion, the quality of local authority assistance can be patchy
- the business and planning advisors may not be suited to the particular scheme
- the developments often need to proceed outside the development plan process
- conflict between the general desire for out of town development accessed by car and Central Government policy to concentrate on modes of transport in urban areas
- resistance from owners of neighbouring properties.

The recent policy shift towards sustainability and reduction of reliance on the car is commendable, but often in direct conflict with the operational process of many clubs, which tends to favour greenfield sites. Developing in urban areas can raise conflict; at the Oval cricket ground, 90–95% non-car access makes for one of the most sustainable grounds in the UK and redevelopment is proving extremely difficult to perform, due to possible impact of change.

Finally, any scheme and ancillary alternative development must be commercially viable. Funding proved to be successful in opening up derelict land and creating a new transport link. In Portsmouth the added retail development contributed to the growth of objections, and ultimately the scheme failed to gain approval following a public enquiry

Where next?

At present, the Government is reappraising the policy framework which deals with stadia development and reviewing the effectiveness of PPG17. Two things should emerge from this:

- a more detailed framework for dealing with stadia which aims to move forward the importance of development plan policy
- confirmed support for the redevelopment of existing sites or use of brown land linked to infrastructure improvements.

As in many changes in property development, the easier schemes have been undertaken. We now look to modern design to achieve the ambitions of sports clubs in the complex and challenging environment of a constrained town planning framework.

HARNESSING TECHNOLOGY FOR THE YEAR 2000

The new stadia at Epsom and Cheltenham for racing, for cricket at Lords, for athletics at Don Valley and Sheffield and for football at Huddersfield are all national examples of successful modern design. Their form inspires and enhances the surrounding area, the design works in an organisational and urban way and long-term aims of the design are realised. While their form inspires, and they are well planned and organised, key to their success is that they work in an urban context, both visually and socially.

To achieve a measure of lasting distinction in stadia operation requires a rigorous and imaginative appraisal of the market place, of asset value and physical layout as well as of operation and maintenance if the facilities are to achieve image, be comfortable and affordable and safe. Location within the 'townscape', view and approach to and from that townscape is as important as the imagery and content of the facility itself.

The local community: providing the infrastructure and inhabiting the space.

The real improvement to sporting infrastructure has to come from the 'lower' divisions by providing sporting and leisure excellence at community level – multi purpose, high class and providing 'engaging' environments within a new generation of buildings. A new facility must give back more than spectator participation to the community by focussing local interest in healthy lifestyles, leisure and youth participation, whatever the sport. Forming a social network around the 'architecture' for the league team or evening racing venue can invigorate community life and create demand for a new type of 'branded' sports facility; providing facilities, not just for sport but for the community as a whole, will appeal to large businesses, wishing to place their 'identity' in association with such activity. And it could be big business.

In his report, Taylor held out the example of the Galgenwaard Stadium in Utrecht (Fig. 1) completed in August 1982 – ostensibly four conventional covered stands located around the sides of the football pitch to give 12,000 seats and 8,000 standing spaces. However, behind each stand, in the corners (often left open) and on three of four sides is some 30,000 m^2 of shops, offices and workshops, while the club's own offices and rooms are on the fourth side.

Fig. 1. The Galgenwaard Stadium in Utrecht.

The surrounding car parks set into landscaping are then used throughout the week as well as for evening weekend sporting fixtures. Parallel financing and parallel land use are all important for economic prosperity and urban sustainability.

By today's more adventurous standards, this design is architecturally rather plain and lacks 'brand' identity but it still provides a fine 'developmental' model for all ranges of sports and leisure. In the same country, the central city stadium for PSV Eindhoven (1988) for 16,000 seats and 11,000 standing places provides a clearly identifiable image, branding the 'Phillips' products, joining a six-storey multi-complex to two-tier high-tech steelwork with closure glazing, with electronic information boards behind. This facility even has gas radiant heaters on the underside of the roofs to warm the spectator on winter days! [2]

Closer to Buro Happold's own experience of 'multi-functioning' is the 30-acre Kowloon Park in Hong Kong with its gardens, birdwalks, and sculpture park. This development includes an Olympic swimming and leisure centre on an 11-acre part of the site that was previously a barracks. Through the innovation of a floating pool floor to join the competition swimming area to the adjacent free-form leisure pool, the naturally-lit 50 m swimming and 15 m diving pool expand to a family leisure bathing facility for four sessions of 1,500 people per day. In summer, a further four sessions of 2,000 swimmers per day can enjoy the cascading outdoor pools within the surrounding landscaped park. Within the complex as a whole are full multi-storey changing facilities, bowling, squash, a 52 × 36 m space for volley ball, etc., shops, two types of restaurant and a 7-a-side football pitch. Also included is a specific screened disabled and warm up 25 m swimming facility beneath the 2,000 spectator seating. Kowloon Park provides the facilities, the leisure and the landscape for 1.5 million people! (Figs 2 and 3) all within a facility that has been meticulously designed on Miesian principles of architecture and carefully optimised principles of engineering structure and services.

Fig. 2. Kowloon Park, Hong Kong.

Fig. 3. Kowloon Park, Hong Kong.

Siting the stadium – where should we be building these new facilities?

While the current emphasis is on development on brownfield sites and in urban areas, there may be the rare instance where a green field site, adjacent to the suburban motorway network provides a leisure and sporting opportunity which includes formal and occasional spectator activity.

The new international cricket academy and ground for Hampshire Cricket Club by Michael Hopkins Architects is such an opportunity (Fig. 4). Supported by Sports Lottery money, the complex includes an adjacent golf club, practice ground and bowling centre. This site, a neighbour to the existing tennis and swimming centre at Eastleigh, has almost direct access off the M27 motorway. The development is formed by reshaping the natural landscape; relatively small, elegant low key buildings housing the various facilities are placed within this terraced earth bowl, which then provides formal seating for 6,000 persons (of which 750–1,000 are covered) and further space for 2,000 unfixed seats. Naturally, on good weather days picnicking on the earth slopes is also possible. The venue is essentially a serious business venture planned as a social as well as sporting focus for the Club. In addition, it aims to provide significant educational, training and medical facilities within the complex.

Normally, improvement will be by imaginative regeneration of existing facilities within the present townscape by generating new business activity. The new 'Teachers' stand for Bath Rugby Club (and others) was designed to create new business by employing 'contemporary' technology. A striking form is achieved by tensile membrane roofs, which, together with longspan steel vireendeel floor trusses and precast concrete slabs enabled a column-free seating area. Fully prefabricated construction packages were assembled to enable completion within the closed season while still creating a new image within an historic city (Fig. 5).

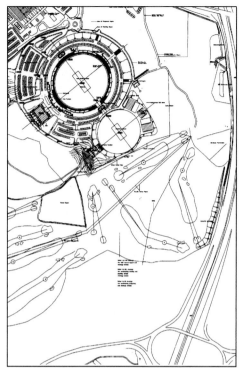

Fig. 4. Hampshire Cricket Club.   **Fig. 5. Bath Rugby Club.**

Land reclamation

PPG17 encourages the use of urban brownfield sites for sporting activity, modern reclamation technology and expertise makes this possible and generally the public transport infrastructure benefits by the enhancement needed. Designs for Manchester 2000 (below) and indeed the Millennium Experience recognise this.

For a high-profile venue such as that proposed by the Balfour Beatty Group for the Manchester 2000 Olympics a tight masterplan for the design of the facilities themselves was necessary. This project, for which Buro Happold were the design engineers, embodied urban regeneration on an old gas works site. A basic network of walkways and cycle ways ensured the primacy of non-car users, supported by good signage and high quality hard and soft landscaping (Figs 6 and 7). The stadium was placed next to the rail and bus station and the development was designed to form a significantly attractive new district. A landmark for the city, it was to contain 46,500 m² of offices, a 250-bed hotel, 41,800 m² of multi-screen cinema and retail areas, centred around the Olympic arena itself. The aim was to provide a multifunctional facility which would create up to 4,300 new jobs. Additionally, the local economy would benefit during the construction by the creation of some 7,500 jobs at a cost of £12,000 per employee.

The stadium itself, with warm up track, a 6,000 seat multipurpose arena and 2,400 car parking spaces, was to be 'state of the art' with 80,000 all seater spaces. The main dilemma in developing the design was the incompatibility between Olympic athletic design and league football.

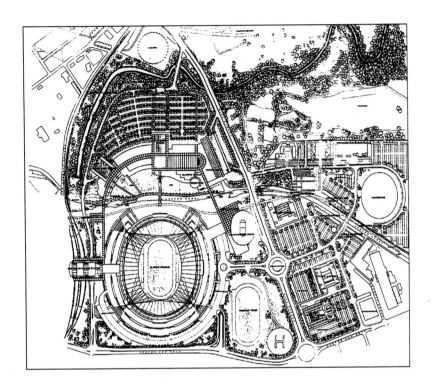

Fig. 6. Site plan, Manchester 2000.

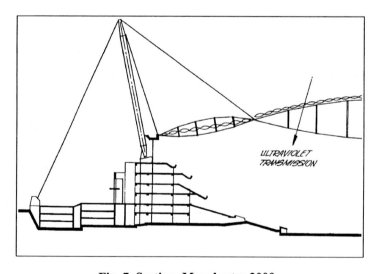

Fig. 7. Section. Manchester 2000.

SAFETY ISSUES

All seater stadia?

After the Taylor report, all football grounds were to become seated, while lower division grounds could only retain terraces if brought up to the highest standard required.. Should we be building all seater stadia for every level and type of sporting fixture as the Taylor report recommended? Certainly we can not for the circulatory events such as racing, rowing, golf. All seater stadia provide more comfortable accommodation and facilitate crowd management every spectator has a numbered place in a stand with a known capacity. However, the lower density of the 'all seater' has the disadvantage of increasing viewing distance, thereby decreasing the intensity of interaction with the event while also requiring a greater capital cost per spectator. This is particularly so in Northern Europe where 'seats' need the cover of large overhanging roofs for protection against rain. Such capital investment seems really only justifiable in a finite economy for the large, prestigious grounds, where high ticket prices are sustainable with well designed modern standing terraces, with properly controlled entrances.

Designing for evacuation

The classical method for determining evacuation time is to apply empirical correlations of crowd flow rates through restrictions and on different terrains such as floors, ramps and stairs. This method, however, requires extension to cater adequately for real cases which include merging people flows, differing mobilities and behavioural responses of individuals in a particular situation.

To design the most effective means of escape, Buro Happold uses computer analysis to predict evacuation patterns for the considerable internal populations of stadia at any moment. The graphical nature of the EXODUS program enables the evacuation to be viewed in progress (Fig. 8) and escape routes to be optimised, while Computation Fluid Dynamics (CFD) enables smoke movement and heat distribution to be predicted (Fig. 9).

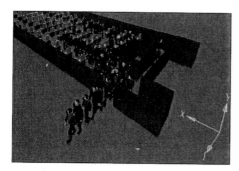

Fig. 8. Simulation of evacuation.

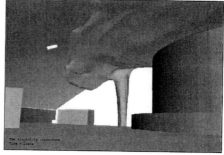

Fig. 9. Computation Fluid Dynamics (CFD) simulation of smoke movement.

EXAMPLES OF GREAT STADIA FOR TODAY AND THE FUTURE

Modern engineering expertise today's stadia to take spectacular forms; examples abroad show how technology can be used to create highly imaginative designs as in, for example, the proposal for a stadium in Jeddah for which Buro Happold were the structural engineers (Fig. 10). Stadia designs in the United Kingdom, however, have tended towards the pedestrian – but there is potential for change. Two examples follow, the first a competition scheme design and the second a current structure which promises to be one of the most significant developments in the UK ever.

Fig. 10. Proposal for King Fahd International Stadium, Jeddah.

Wonderworld
Natural lighting of daytime sporting events has always been prized, especially as this will sustain the growth of natural grass. The proposed use of innovative ETFE transparent and insulative foil cushions for the 20,000 seat arena at Wonderworld, Corby foreshadowed developments in this area (Fig. 11). The comfort and flexibility of a closed arena available for a range of sporting, conference and exhibition events in protected daylight can then be provided, together with the pleasures of natural grass, particularly when the growing tray can be moved to an outside environment to reveal the arena floor beneath, as is now possible. Certainly, these transparent foil cushions, as used for the many leisure pools for Centre Parks, have provided an inspiring environment for the new Tennis Club at Eastleigh, Hampshire (Fig. 12).

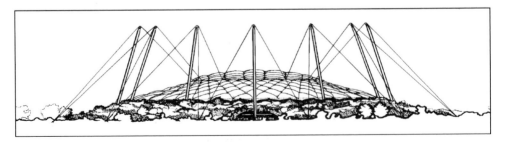

Fig. 11. Wonderworld, Corby.

Fig. 12. Tennis Club at Eastleigh, Hampshire.

The Millennium Experience

The Millennium Experience addresses the typical issues for a major sporting and leisure venue. The technologies of longspan lightweight structures, of land reclamation, of large scale environmental servicing are all used, combined with ready access to the extended river, bus and railway networks, all on a site reclaimed from a disused gas works (Fig. 13).

The envelope for the Dome arena is constructed with the help of computer driven precise prefabrication techniques. Galvanised post tensioned cable nets and integrated patterned tensile roofing membranes of ptfe coated glass fabrics make up and support a double layer of tensile roofing, designed to provide a measure of translucency (7%), as well as protection against condensation on cold winter nights.

The envelope is designed to create an environment comfortable in all seasons to all occupants whose internal climate at the level of occupation is determined by computational fluid dynamics (CFD) as discussed earlier and used for fire safety engineering. It is particularly appropriate to large volumes such as stadia, where the assumptions made in simpler flow models are invalid. The use of CFD can also be extended to smoke movement in a large (covered) stadium to predict the development of a medium growth (15 MW fire) inside the 370 m diameter 50 m high dome, as part of evacuation modelling.

SUMMARY

As both 'landmark' developments and gathering places, stadia have great urban potential. Now, more than ever, there is the opportunity for owners and managers to exploit the 'brand' opportunities of new leisure and lifestyle markets in addition to spectator attendances – a highly desirable demographic audience to marketeers. Good planning, visualisation and design, linked to modern design technologies, video and sound systems and marketing strategies can be tailored to enable construction of radical concepts in sports facilities whose commercial viability is then also ensured.

Fig. 13. Millennium Experience

REFERENCES

1. Home Office. *Final Report, The Hillsborough Stadium Disaster; 15 April 1989.* HMSO, London, 1989.
2. *Appraisal of Sports Grounds,* The Institution of Structural Engineers, May 1991.
3. Simon Inglis, *The Football Grounds of Europe,* Collins Willow, London, 1990.

2 MAKING A NEW STADIUM SPECIAL

E. A. R. LENCZNER
Ove Arup & Partners, London, UK

SUMMARY: This paper argues that major new stadia projects should regarded as important public buildings and, accordingly, that their design should be given sufficient attention that they will be seen as valued buildings within a city. It goes on to explain the importance of considering design aspects beyond the purely functional and attempts to identify those aspects of design which can make new stadia 'special'.

Keywords: American stadia, city developments, dome stadia, football clubs, history, internal and external appearance, roofs, seating bowls, stadia as public buildings.

INTRODUCTION

For sports enthusiasts, stadia provide the backdrop to memories of past glories and evoke fantasies future glories. Titles such as 'Theatre of Dreams' and 'Venue of Legends', which Old Trafford in Manchester and the existing Wembley Stadium have been given respectively, illustrate the special role these stadia have in the mind of those who congregate within them with anticipation of spectacle and excitement. In a sense, major football stadia can be considered as opera houses for the masses. This seems all the more appropriate now that stadia are regularly used for music events as well as sporting. As such, stadia should be given the architectural treatment such a status suggests. People expect the design and ambience of a great theatre to heighten the senses of attendant spectators and so if the stadium is to be considered as a very large theatre so it should be designed to heighten the senses of all those who visit it.

The 'organic' evolution of the typical British football ground has meant that most football clubs still play at old grounds dating back several generations. There is now an impetus to build a new generation of stadia to replace many of these old stadia and spectators can look forward to higher levels of comfort and amenity than they have previously known. However, in replacing old stadia which were short in comfort and safety, we need to recognise that many of them have a charm such that they have became special places in the hearts of local fans. Many football fans will mourn the passing of their old characterful stadia when they are demolished in to make way for a new ones. It would be a pity if we were unable to design our new stadia to be as 'loveable' as their predecessors came to be despite their obvious shortcomings.

Stadia, Arenas and Grandstands, edited by P.D. Thompson, J.J.A. Tolloczko and J.N. Clarke.
Published in 1998 by E & FN Spon, 11 New Fetter Lane, London EC4P 4EE, UK.   ISBN: 0 419 24040 3

Looking to the future, it is significant that television rights seem destined to play an increasingly important role in the business of sport. It is quite possible to imagine that television companies will not only influence when events are staged but also where. This being the case, it is all the more important that stadia should be attractive to television audiences as well as live spectators.

In designing new stadia built to replace old existing stadia, careful consideration therefore needs to be given to those design issues beyond the purely functional or operational (such as sightline and safety criteria). By doing so will make the difference between the completed stadium project ending up being perceived as just an adequate functional stadium or as a venue with character and ambience to inspire sporting dreams.

HISTORICAL BACKGROUND OF BRITISH STADIA

In Britain, football grounds started to emerge during the end of the 19th century as football clubs became established. Usually these grounds started life as a humble collections of shacks around a field catering for only small crowds. During the course of the 20th century, as football gained more and more popularity, these grounds went on to increase ground capacity through piecemeal development and evolved to become the archetypal football grounds we know today.

Most long-established football grounds were built within the working class parts of cities and for most of its history football has been an essentially working class sport played in front of flat-capped masses. In the same way as factory chimneys and smoke would represent the workplace, the floodlight masts and large shed roofs of the football stadium became symbols to beckon football supporters to the ground on match days. The football stadium came to be the most popular communal forum for townsfolk to gather on a regular basis. Today most British league clubs still play in stadia which they own and which have histories stretching back many generations. This continuity of a football club in a single location means that supporters and players feel a sense of loyalty to the club's history and traditions. These are things which cannot be artificially manufactured and which help explain why even semi-dilapidated old grounds can still be regarded as special places in the hearts of sports fans, even if they offer little physical comfort or safety.

Until the 1990's there was little pressure for football clubs to contemplate major ground redevelopments. The tragedies at Bradford and Hillsborough, where many lives were lost due to poor stadium safety, changed all that. The Taylor report, published in 1989, not only obliged grounds to be rebuilt to meet new safety standards, but seemed to act as a catalyst for change in football in Britain as a whole. Over a few years following the Taylor report, football broadened its appeal to become of interest to all classes and all the family. Clubs gained sponsorships from multinational corporations and new income from television rights. Suddenly football in Britain has gained a broad appeal and is now a big money business able to invest in new stadia.

With more and more clubs considering plans to build new stadia for the next century, a responsibility falls upon stadium owners and designers to ensure that, whilst meeting the levels of comfort and safety now expected, the new generation of stadia are able to engender that special feeling in the same way as many of the previous generation of stadia managed to do.

THE STADIUM WITHIN THE CITY

Today, a major new stadium is likely to be one of the largest single mass buildings within a city's urban fabric and will represent a popular city forum regularly attracting the city's largest assemblies of people. It will be one of the most significant new buildings to be built and should be considered as a worthy and prestigious development for insertion into a city centre setting. A parallel can be drawn with the way in which major railway stations were introduced into city centres in the 19th century, and which are of comparable size to a stadium. Today these railway stations are accepted as rightfully belonging to the city centre fabric; in some cases they almost define the centre of a city. In the future new stadia could be similarly accepted as belonging to the centres of our cities.

Historically, stadia and large amphitheatres have been regarded by societies throughout the world as important public buildings within the city. Both the Greek and Roman civilisations, for example, built their stadia as great municipal theatres at strategic positions within their cities. Where new stadia are proposed in Britain today, the trend is for them to be located remote from city centres on edge-of-town sites. This seems unfortunate as it fails to properly recognise the civic importance of a stadium. It is understandable that such sites are cheaper to build on than city centre sites, but by building new stadia in the same sort of development land as warehouses is to ignore the potential benefits of placing them close to the heart of a city's urban fabric.

For the next century there seems to be an opportunity for stadia to provide once again a focal point in city centres. This may become particularly important as many city centres may tend become run-down as edge-of-town shopping malls steel much of their previous core activities. Arguably, city authorities should invest as partners in stadia projects to ensure that they are given the status they deserve within the modern city and so benefit from the economic spin-off they will generate for the city centre. By placing stadia within city centre districts, they will be well served by the city's established public transport system. Generally speaking, peak demand for the public transport generated by the stadium would not coincide with the rush hour traffic from the city centre's business district. Because they are generally well served by public transport, city centre sites have less need for car parking provision than suburban sites.

The location of a stadium is likely to effect its architectural treatment. Today, where stadia are built to low budgets on sites typically occupied by light industrial development, they are likely to end up with the same banal architectural treatment as their warehouse neighbours. This diminishes the standing of the stadium as a building type both in terms of its location and appearance. This is in total contrast to the Colosseum in Rome whose location and architecture indicate that the stadium was given a high importance within the city.

Examples can be found where major stadia are located in the heart of a city. In Cardiff the National Stadium (more popularly known as the Arms Park) has stood on the bank of the River Taff amidst the city's central business district for as long as people can remember. By building the new Millennium Stadium on the same site the stadium should remain an important focal point of the city centre. Other examples where stadia feature as part a city's central areas are in Madrid, where the impressive Santiago Bernabeu Stadium stands proudly on the city's prestigious Castellana thoroughfare, and in Toronto, where the Skydome forms an important part of the city's downtown silhouette next to the famous CN Tower.

Clearly, in some historical cities, there may be cases where for architectural preservation reasons it would prove difficult to find a suitable central site. However, the closer a stadium can be placed to a city's heart the better the chance of the stadium becoming special in the heart of its citizens.

LESSONS FROM RECENT AMERICAN STADIUM DESIGNS

In exploring how new British stadia should be designed, it is instructive to see how new stadia have developed in North America since the 1960s. During the 1960s and 1970s new stadia in the United States tended to be built on open, edge of city sites allowing the new stadium to be surrounded by vast expanses of car parking which was provided to meet the expectations of a public who lived and breathed the automobile. These stadia were often designed as combined sports use stadia which would be convertible between baseball and American football use. During the same period some of the new stadia were built as 'dome' stadia totally enclosed by a fixed roof and with artificial grass fields.

By the 1990s new stadia were being designed dedicated for baseball or American football use and many of these were destined to supersede stadia built only 30 years previously for combined use. Two reasons can explain this reversal. First, baseball fans found the combined stadia lacked the character of a 'real' baseball park with the classic horse-shoe grandstand configuration and open outfield. Second, major league baseball and American football clubs each became economically strong enough in their own right to justify building separate stadia.

One of the first new generation baseball parks to be built was the Oriole Baseball Stadium in Baltimore. Whilst incorporating the type of up-to-date amenities to meet the demands of the modern day spectator the architecture of the stadium reintroduced some of the classic baseball park vernacular, such as exposed steel framework and an asymmetric seating bowl, which help evoke a sense of the game's traditions. Other new baseball parks at Cleveland (Jacob's Field) and Atlanta (Turner Field) also feature similar 'retro' style. At Baltimore, the design of the stadium was also effectively integrated into the urban fabric of the city's downtown district with views of an old warehouse featured in the stadium planning.

It seems that if a new stadium fails to provide a special enough character, paying spectators may begin to wish that they could have their old stadium back. In the United States stadium designers have responded to baseball nostalgia in the newer ballparks in an almost theme park style. Good quality stadium architecture need not, however, rely on a 'sports heritage' design style to provide the stadium with appropriate character. It is quite possible for stadia to be designed in a modern style but still include features to endear the stadium to its visitors.

Baseball clubs using 'dome' stadia now tend to want to move to stadia with natural grass fields open to the sky allowing the game to be played in its traditional outdoor setting. Despite the advantage of being weatherproof which dome stadia offer, both spectators and players alike seem to prefer the baseball and football games to be played on a natural grass pitch in an outdoor environment. Most new stadia projects in the United States, for either baseball or football, are now designed as either open air stadia or else with retractable overall roofs.

Attempts to design a stadium to cater for as many sports as possible may sometimes backfire as the stadium may become 'Jack-of-all-trades-master-of- none'. The Americans have found that baseball just isn't quite the same when played in what is essentially a football stadium. In Atlanta, the new baseball stadium for the Atlanta Braves has been built to replace the adjacent Fulton County Stadium which was built as recently as 1965 for both baseball and football use. As the Atlanta Falcons football team had already moved to a new home in the Georgia dome a couple of years previously, the Fulton County Stadium became obsolete and has now been demolished. Outside America, the most common example of two incompatible sports sharing a stadium is where athletics tracks are placed around a football field. This normally means that spectators for the more popular sport, football, must put up with excessive viewing distances. If football clubs in Britain become financially strong enough, there seems no reason why the American example of building separate stadia for different sports needs should not be followed.

Whilst comparisons between Britain with North America may have limited validity because of cultural and economic differences, the American experience provides warning of potential pitfalls when old-established stadia are replaced by new stadia. Essentially, if the design of a new stadium compares poorly with event owners' or spectators' aspirations it may become obsolete before its design life is reached.

EXTERNAL APPEARANCE OF A STADIUM

A new major football stadia as a single development represents a very large building in urban terms. If it is to be integrated successfully into an urban fabric, a stadium's frontages to the surrounding streets need to be designed as architectural building facades rather than presenting a crude enclosure to the back of the stand.

The comparison between a stadium and a major railway station is useful again. Railway stations were built as 'cathedrals of the modern age' in the 19th century and had their exterior facades express their importance to make them unmistakeable as to what they were. Similarly a stadium should appear to stand proud in its setting and to invite spectators in rather than appear as an impenetrable fortress.

Because stadia tend to be designed from their inside to their outside, they can often appear as introverted buildings whose back is turned to the street. Here a lesson can be learned from some of the great opera houses or theatres which make a point of presenting an appealing extrovert facade to the street. By incorporating activities into the stadium's external frontage it becomes easier to make the stadium seem animated when viewed from the outside. However it is also possible to arrange the stadium's necessary concourse spaces and vertical circulation routes such that they form an interesting visual composition.

Many of the older European stadia were designed with towers to provide a strong identifiable image for the stadium within the city. Examples of such stadia include the Olympic Stadium in Helsinki and the Stadio Dall'Ara in Bologna. In England, the twin towers of the old Wembley Stadium have come to be world famous icons which sports fans can immediately identify.

A distinctive external image of a stadium can also be achieved through the expression of its structure. In particular, the geometric form of a stadium's roof structure can also be used to provide drama to a stadium's appearance. This is especially true if it is designed as a unified system rather than having independent structures for stands on each side of the stadium.

The Munich Olympic Stadium's distinctive profile is provided by its impressive tensile roof structure featuring a series of masts. At Bari in the south of Italy, the San Nicola Stadium has a distinctive exterior appearance due to its expressive concrete stand superstructure. Since its construction for the World Cup finals in 1990, it is probably no exaggeration to say that the stadium has become Bari's best-known piece of architecture.

Each new stadium needs to establish its own identity to distinguish it from others. It is the stadium designer's responsibility to create the identity with which local fans feel they can associate themselves and by which towns or cities become known.

INTERNAL APPEARANCE OF A STADIUM

The most important part of a stadium which can makes it feel special is its interior. As with a well-designed theatre, to enter into the inside of a well-designed stadium should cause a sense of exhilaration, even when the stadium is empty. Various devices can be used to add character and ambience to a stadium. Some important but not strictly functional design issues affecting the quality of a stadium interior are identified here.

Stadia with seating bowls which have stands or tiers with a variable height around the arena tend to be visually more interesting than those where there is a constant profile all the way around their arenas. Variation can be achieved in more than one way and can correspond to the patterns of preferred seating positions around the arena. Variation in the heights of stands can be made by changing the heights of stands between stands on different sides of the arena. Alternatively, curved elevational profiles can be used to define a gradual variation in the height of stands around the arena.

Seating bowls also tend to be more interesting when they have some degree of asymmetry. This also helps spectators feel a better sense of orientation within the stadium. A good example of a stadium with both a variable height stand profile and asymmetry is the Nou Camp Stadium in Barcelona.

The intimacy of a stadium can be improved in two important ways. Firstly spectators should be placed as close as practical to the arena. Stadia with long distances between the sports field and the first row of spectators, typically due to the presence of a running track, tend to lack the intimate ambience of those where spectators are positioned tightly around the arena. A modest size stadium with a close-packed seating bowl can sometimes generate more atmosphere than one twice its size whose spectators are too remote from the arena. Secondly, stands with seating rows laid out following a curved plan mean that each spectator is able to see other spectators to each side within the same stand and this engenders a better sense of intimate communion within the crowd. Even a small curvature can make the difference. In Spain there are many football stadia which feature such gentle curves in the seating rows, perhaps this is because the local people know the intimate feel of the bullring.

From within, the stadium stadia can be given more character if distinctive visual reference points exist around the seating bowl. Such reference points can be in the form of towers or even floodlight supports. Flags can also be used to animate the stadium interior. Sometimes an unusual quirk or 'accident' such as an odd arrangement of tiers or box accommodation can act as a useful device to made a stadium distinctive.

When a stadium is set within a location with surroundings rising above the back edge of a stand, interesting visual links can be made between the inside and outside of the stadium. If this is possible it can be a useful device to remind spectators where they are and affirm the sense of the stadium belonging to its site. Old pictures of English stadia with trains running past the back of a stand give a strong sense of its location. In modern stadia it may be possible to snatch glimpses of landmark buildings or even hills around the stadium to achieve a visual link from inside to outside the stadium.

The careful use of colour within a stadium can sometimes make an ordinary stadium seem good. To emerge into a stadium seating bowl through a vomitory, the visual impact of the grass arena should be complimented and enhanced by the colour of seating and the stadium's building fabric. As elsewhere however, the exaggerated use of colour can sometimes have a detrimental rather than beneficial visual effect.

The nature of a stadium's roof has one of the biggest impacts on the appearance of the interior of a stadium. Generally speaking, the more that a large roof allows daylight through onto the spectator areas beneath, the better the stadium ambience is likely to be. At one extreme a stadium roof could appear as a rudimentary heavy 'lid' over the spectators, whilst at the other it could appear to float elegantly as a translucent canopy. To illustrate the difference one could compare the new roof over the Gottleib Daimler Stadium at Stuttgart with the roof over the Prater Stadium in Vienna. Both roofs have a similar tensile structural system to support them but the main difference being that, whereas the Vienna roof is essentially opaque with sheet metal cladding, the Stuttgart roof has a translucent membrane throughout.

The geometric form of a large stadium roof, as well as playing a potentially dramatic role on the composition of a stadium interior, will affect the cast shadow patterns on the arena. In some cases the roof shadow will be so great that most of the grass pitch will be in shade for most of the time. Not only does this have potential consequences on grass growth within the stadium, it also changes the environment in which games are played. For example, since the new West stand at Twickenham was completed, most of the Five Nations international rugby matches played there are almost completely in the shade whereas before they were not. For field sports, we need to decide if it matters, in ambience terms if nothing else, that games in the future might never be played in sunlight. The English FA Cup final might not be quite the same, for example, if the match was always played in shade rather than sunlight.

Now that retractable roofs are being considered for more new stadia projects perhaps retractable systems should be devised not only to fully enclose stadia but also allow the roof opening to be enlarged to let in more sunlight during fine weather.

CONCLUSION

Over the next decade or so the surge of new stadia developments seems likely to continue to meet the demand for higher levels of comfort and amenity by both spectators and event owners. There is a risk however that in design of these new stadia, although meeting the strict functional requirements of a brief, they may lack the quality that a major public building in a city should be expected to have. Stadium owners and designers should be encouraged to give careful consideration to the non-functional aspects of stadium design, such as those mentioned in this paper, which can make a new stadium a special and valued highly by all those who come to know it.

3 THE NEED FOR A NEW APPROACH TO INITIAL COST ESTIMATING FOR STADIA AND GRANDSTAND FACILITIES

J. A. COXETER-SMITH and J. D. WOODROUGH
Davis Langdon & Everest, London, UK

SUMMARY: This paper demonstrates the limitations of the traditional, *cost per seat* approach to initial cost estimating for stadia and grandstands and sets out a new approach which is sensitive to the key variables affecting costs thereby allowing greater realism and accuracy.
Keywords: Accuracy, *cost per seat*, estimating, initial cost estimates, key variables, limitations of historical approach, realism.

INTRODUCTION

Stadia, grandstands and other developments for spectator sports form a significant and growing slice of the construction market. In football alone, many clubs have commissioned substantial stadia developments following the issue of the Taylor Report [1], the increasing popularity of football requiring larger capacities than that provided by purely fixing seats to existing terraces, and the growing financial incentives of new, larger facilities. In the past 10 years 12 of the 92 Football Association Premiership and Football League clubs have moved to all new stadia with an aggregate capacity of 236,600 [2]. To this can be added the numerous developments for other sports such as the Rugby Football Union stadium at Twickenham and the plethora of grandstands provided for professional club rugby, cricket, horseracing and lower league football clubs.

With further stadia developments in the planning or construction phases including the Millennium Stadium (Cardiff), the National Stadium (Wembley) and the National Stadium (Manchester), this volume shows every sign of continuing, and quite probably increasing over the coming years. Some construction industry analysts have estimated that this sector of the UK market will be worth more than £1bn over the next five years [3].

The main sources of finance for these developments include stockmarket flotations, professional investors, television revenues and lottery funds. Investment funding is attracted by the levels of return which are achievable, but the increasing involvement of the City in this sector does place greater demands on the stadia developer/owner/operator in terms of the expectations as to the calibre and degree of professionalism of financial management. Such expectations extend to the area of the financial appraisal of potential developments at the feasibility stage and thus a realistic estimate of the capital cost of the proposed scheme is essential.

Stadia, Arenas and Grandstands, edited by P.D. Thompson, J.J.A. Tolloczko and J.N. Clarke.
Published in 1998 by E & FN Spon, 11 New Fetter Lane, London EC4P 4EE, UK.   ISBN: 0 419 24040 3

THE NEW APPROACH

Introduction

The historical approach to estimating the cost of stadia or grandstands has been based on the *cost per seat* method. Here a cost per spectator is applied to the planned capacity to give an estimated cost. The limitations of this approach are shown in Table 1.

Table 1. Comparison between published costs and costs estimated by the *cost per seat* method.

Stadium	Capacity	Estimated costs		Published costs	
		£/seat	Total	£/seat	Total
Riverside Stadium, Middlesborough FC [3]	30,000	-	-	400	12,000,000
Grandstand, Lords Cricket Ground [4]	6,000	400	2,400,000	2,166	13,000,000
Reebok Stadium, Bolton FC [5]	25,000	400	10,000,000	1,232	30,800,000
Tattersalls Stand, Cheltenham Racecourse [6]	3,220	400	1,288,000	3,105	10,000,000

Note: Costs not adjusted for location or inflation.

Table 1 shows the discrepancies between estimated and actual costs which would arise from the application of a cost per seat, derived from published data, on one stadium development in cost estimates for several others.

We can deduce from Table 1 that the costs of stadia and grandstands vary from development to development. In response we have developed an approach to cost estimating which recognises the factors which may cause costs to vary and is sensitive to changes in the balance of these factors.

Key variables

Peter Smith, a partner in stadium architects, Miller Partnership is quoted as saying "The key elements of new stadiums are safety, sightlines, spectator comfort and movement, cost effectiveness, architectural expression and commercial potential" [7]. Similarly, through experience and analysis, we have compiled our own list of the key elements or variables which are significant in shaping the costs of stadia or grandstands.

The key variables need to be determined in the earliest stages of the project if cost estimates are to be realistic. They need to be thought through carefully and may be applied in a series of alternative scenarios to illustrate the budgetary implications. The following list highlights some of the key variables and a number of implications relevant to that variable.

- *Location*: A new stand built to replace an existing stand will mean demolition costs are inevitable, may result in restricted access due to surrounding stands and buildings, and a compressed programme to minimise disruption to the business. Whether a stand is situated to the side or at the end of the pitch has implications on spanning distances affecting quantity and complexity of roof and frame structures. A greenfield location will mean infrastructure costs, ground conditions and the land purchase costs need consideration. Geographic location is also relevant as construction costs vary from region to region and country to country.

- *Infrastructure*: New or improved spectator transport facilities may be required. Revised capacities may create the need for extra car parking space or improved public transport. If a greenfield site is being utilised a complete new access road and parking facilities are likely to be essential.
- *Tier arrangement*: The number of tiers required will be dependant on capacities required, footprint available, planning parameters, sightlines and maximum viewing distances. For example – a second tier of seating will result in increased structural frame complexity and quantity, greater loadings on foundations, the need for vertical circulation, more complex escape/safety matters, increased floor area, and a larger elevation to clad, all of which have major cost significance.
- *Type of roof*: Another major variable in which choices may vary from open stands with no roof to a fully covered stadium with a retractable roof. Unobstructed views, as recommended in The Taylor Report, means column free viewing and therefore greater complexity of the roof framework. The roof may also be seen as the prime object in which an 'architectural statement' can be achieved. If this is the case aesthetic enhancements to the visible structure or the coverings may result, once again affecting the cost.
- *Gross floor area*: Table 2 shows the kind of differences which can arise. Football stands in the examples range from 0.71 m²/seat to 1.88 m²/seat – a 165% difference in area per spectator. Gross floor area is affected by tread depths, seat widths, vomitory spacing, etc.; whether or not player, staff and office facilities are required (generally in a main stand); the number and size of basic facilities such as bars, kiosks and WCs; and, importantly, the extent of hospitality and retail space to be provided. The sport for which the stand is designed will affect the floor area and facilities within the stand. The stay at a horse racing meeting may last for five hours compared to the 90-minute duration of a football match with the attendant demand for facilities that duration of stay creates. Linked to this is the proportion of spectator capacity to overall stand capacity. A football stand's capacity corresponds to the spectator capacity. A racing grandstand however may only hold 2,000 seated and standing spectators, but may include facilities for the further 15,000 who watch the racing from the rails, betting ring or elsewhere on the course.
- *Quality of finishes*: Finishes both internally and externally affect cost. A general assessment of the quality level is required, although full details may not be available during the early stages. For example, aspirations may be for frameless glazing to the rear elevation although budgetary constraints or financial returns militate towards a more basic treatment.
- *Fitting-out*: Largely dependant on the facilities available in the stand. Fitting-out costs would be relatively low for a stand containing a concourse and a few bars and WCs. The level of fitting-out will grow as more facilities are included, for example, hospitality boxes will require kitchens and equipment, tables and chairs, televisions, etc.

- *Other facilities*: A gymnasium, sports hall, swimming pool or under pitch basement car parking will have a major impact on the cost of the stand, significantly increasing the gross floor area. Other major items to be considered which do not affect the gross floor area include floodlight requirements, television/operational facilities, large video screens, advertising display facilities, etc.

The list is not exhaustive and the variables are not necessarily independent of one another. In fact the variables may act in conjunction with each other and all may affect common elements. For example, the obvious variables which affect frame quantity, complexity and cost are the tier arrangement and roof type. However, if a large, column free restaurant is required, the hospitality (gross floor area) variable will also have a major impact on beam spans and column sizes, etc.

Table 2. Examples of gross floor areas per spectator

Sport	Venue	Gross floor area per spectator (m²)
Football	Kilner Bank Stand, Huddersfield (Architect - Lobb Sports Architecture)	0.71
Football	Proposed South Stand, Southampton Community Stadium (Architect - Atherden Fuller & Hampshire County Council)	0.72
Football	North Stand, Chelsea, London (Architect - Lobb Sports Architecture)	0.84
Football	North Stand, Arsenal, London (Architect - Lobb Sports Architecture)	0.97
Football	Lawrence Batley Stand, Huddersfield (Architect - Lobb Sports Architecture)	1.28
Football	Proposed East Stand, Southampton Community Stadium (Architect - Atherden Fuller & Hampshire County Council)	1.88
Cricket	Compton and Edrich Stand, Lords, London (Architect - Michael Hopkins and Partners)	0.49
Cricket	Mound Stand, Lords, London (Architect - Michael Hopkins and Partners)	1.03
Rugby Union	South Stand, Twickenham, London (Architect - Lobb Sports Architecture)	1.01
Rugby Union	The Teachers Stand, Bath (Architect - Lobb Sports Architecture)	1.51
Horseracing	Grandstand, Kempton Park, Sunbury (Architect - Lobb Sports Architecture)	2.02
Horseracing	Proposed Grandstand, Rowley Mile Course, Newmarket (Architect - The Goddard Wybor Practice)	2.33
Horseracing	Tattersalls Stand, Cheltenham (Architect - Lobb Sports Architecture)	2.58

WORKED EXAMPLE

An application of the new approach demonstrating the significance of the key variables on the cost of a scheme is illustrated below.

A single tier football stand with a traditional cantilever roof, a gross floor area in the order of $0.70 - 0.80m^2$ per seat (implying only basic facilities such as bars, kiosks and WCs are present; and no hospitality or retail areas of any significance), basic internal and external finishes,on a greenfield site with no programme restrictions, and located in South East England, would cost in the order of £600/m².

This is derived from analysis of a number of new stadia and stands such as the Cellnet Riverside Stadium, Middlesborough (Architect – Miller Partnership), The Stadium of Light, Sunderland (Architects – Miller Partnership and Taylor Tulip) and The John Smiths Kilner Bank Stand, McAlpine Stadium, Huddersfield (Architect – Lobb Sports Architecture).

Table 3 shows the adjustments necessary to reflect a greater complexity of the variables. A two tier stand, with hospitality and retail amenities, player and administration facilities, built on contaminated land, and requiring new infrastructure, such as the Lawrence Batley (formerly Riverside) Stand, Huddersfield can be expected to cost over twice as much as the 'basic stand', approximately £1200 – £1300/m².

The variables are further illustrated in Fig. 1, a cross-section through the Lawrence Batley Stand. It shows the areas allocated to conference facilities, hospitality boxes, vertical circulation in the form of lifts and stairs, facilities for players and staff, retail and entrance lobbies. It is reproduced by kind permission of Lobb Sports Architecture.

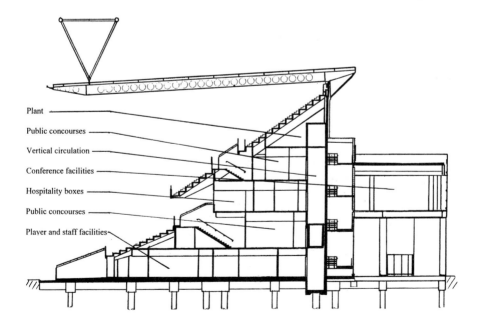

Fig. 1. Cross-section through the Lawrence Batley Stand, Huddersfield.
Architect – Lobb Sports Architecture.

The new approach and the factors (derived from DLE's cost database) are not intended as a tool for cost planning but rather to allow some reflection of the major design issues during the inception and feasibilty phases of a project.

Table 3. Indication of cost implications of key variables.

Key variable	Multiplication factor	Implication
Location	0.90	Geographical location – Yorkshire
	1.02	Allowance for contaminated land
Infrastructure	1.20	New access roads required to reach the disused site and extensive new car parking
Tier arrangement	1.25	Two tiers resulting in increased frame complexity and quantity and larger external wall quantity
Type of roof	1.05	'Banana Truss' uses basic materials, and makes a large architectural impact but circular steel sections in a curved truss have cost implications
Gross floor area	1.15	Corporate and retail facilities including a conference centre and hospitality boxes
	1.10	Player, administration and sundry facilities
Quality of finishes	1.05	Good quality finishes to part of elevation
	1.02	Internal finishes to hospitality areas
Fitting out	1.05	Corporate and retail facilities
	1.02	Player, administration and sundry facilities
Other facilities	-	None included in scheme. Floodlights excluded from this example as they are on stand alone towers and therefore part of the overall stadium cost
Compounded factor	2.10	

CONCLUSION

Stadia, grandstands and other developments for spectator sports form a significant and growing slice of the construction market. Analysts have estimated that the value may exceed £1bn over the next five years. Accuracy in financial appraisals of the feasibilty of schemes is essential in order to satisfy the scrutiny of City financiers and other professional investors.

The historical approach to estimating the cost of stadia or grandstands, the *cost per seat* method, has severe limitations and is unlikely to meet these demands. The consequences of an inadequate or insufficiently robust financial appraisal might include the failure to obtain funding from these sources.

We have developed a new approach which allows for the recognition of the key cost determining variables in preparing initial cost estimates thereby addressing the criticisms levelled against the historical approach and allowing greater realism and accuracy. The new approach is demonstrated in this paper.

We are now working towards the refinement of this new approach which will be furthered as it is tested more widely and the database amplified with the analysis of even more projects.

REFERENCES

1. The Rt. Hon. Lord Justice Taylor. *The Hillsborough Stadium Disaster . 15 April 1989. Final Report.* HMSO, London, January 1990.
2. Ley, J. Sunderland entering new era. *The Daily Telegraph*, 30 July 1997. pp. 38.
3. Cook, A. Multimillion pound transfers. *Building*, 14 March 1997. pp. 54–8.
4. Baillieu, A. News, the week. *Architects Journal*, 21 December 1995. pp. 7.
5. Binney, M. Temples to team spirits. *The Times*, 2 September 1997. pp. 37.
6. A stand for all seasons. *Steel Design*, Autumn 1997. pp. 8–9.
7. Binney, M. Temples to team spirits. *The Times*, 2 September 1997. pp. 37.

4 OLD STADIA MEET CURRENT REQUIREMENTS BY RENOVATION, REHABILITATION AND REPAIR

N. S. ANDERSON and D. F. MEINHEIT
Wiss, Janney, Elstner Associates, Inc., Northbrook, Illinois, USA

SUMMARY: This paper discusses case studies on vintage cast-in-place reinforced concrete stadiums that are functionally obsolete, yet have been renovated to make the primary structural system serviceable. Some structural repair schemes for rejuvenating old stadiums will be described along with the typical deterioration problems encountered causing repairs to be initiated. New, modern amenities can be added to older stadiums provided repairs, inspection monitoring during the future service life, and maintenance are considered as part of the continued use of an older facility. The authors will review three case studies of reinforced concrete stadiums built in the 1920s and are being made functional for use in the 21st Century. These cases are excellent examples of situations that had been neglected, from a maintenance viewpoint, for many years but have been repaired and can expect to be functional for an additional 30 to 50 years.
Keywords: American Football, college, concrete, deterioration, rehabilitation, stadia.

INTRODUCTION

Professional athletic teams in the United States are demanding new stadiums and arenas containing amenities that currently do not exist in older facilities. New facilities require substantial investments by private owner groups, or the new facilities are built with municipally funded programs.

College athletic facilities are also competing for 'market share' attention. This attention enhances the general undergraduate athletic programs and can help finance new campus facilities. Major college (American) football teams cannot command the same new facilities of their professional sport counterparts. Consequently, many colleges must accept their existing facilities, and renovate or rehabilitate their structures to provide a modern ambiance for college athletic events. In the past, renovation and rehabilitation has often been undertaken without considering repair or maintenance of the primary structural system. Repair work on a stadium facility does not sell any seats: amenities in the facility such as skyboxes, improved seating, and concessions are the revenue sources. Accordingly, maintenance funds do not readily exist to repair older stadiums.

Location or setting is another issue with collegiate stadiums. Many existing facilities were built in the 1920s or 1930s in locations away from the main campus buildings. Most major college campuses have greatly expanded since their stadiums were first constructed,

Stadia, Arenas and Grandstands, edited by P.D. Thompson, J.J.A. Tolloczko and J.N. Clarke.
Published in 1998 by E & FN Spon, 11 New Fetter Lane, London EC4P 4EE, UK.   ISBN: 0 419 24040 3

whereby campus buildings now surround these older athletic facilities. Their current locale, within the main college campus, enhances the collegiate sport atmosphere and has the potential of drawing more spectators from the general undergraduate student population. Building new facilities further away from the central campus is undesirable because the athletics program can become 'disassociated' from the university population and undergraduate student support can erode due partly to the lack of convenient public transportation. Therefore, an older athletic facility is usually ideally located, making rehabilitation an even more desirable option.

Similar situations exist regarding the facility location for professional sports teams in major metropolitan areas, although to a lesser degree. Recent professional sports stadia or arenas built in the US have tended to use major tracts of abandoned land or blighted urban areas for construction. Funds for constructing new professional sport facilities are more readily acquired via use taxes.

FACILITY TYPE

Several (American) football stadiums were constructed in the 1920s and 1930s as memorials to World War I soldiers or as Great Depression era social projects by the Works Progress Administration (WPA). Cast-in-place, reinforced concrete superstructures/ frameworks topped with concrete treads and risers characterize a number of these stadia. Generally, the stand structures are either horseshoe-shaped bowls enclosed below the tread and riser system, or two separate structures located on opposite sides of the field sidelines. Again, the underside area is usually enclosed by reinforced concrete walls often having the exterior face containing intricate and detailed architectural features. To accommodate more seating capacity in later years, the bowl ends, in many facilities, were usually filled-in with additional seating.

Because of their construction vintage, these stadia predate modern building codes and have generally been viewed throughout the years as maintenance free structures. Similarly, as these facilities are used only 6 to 10 times per year, it can be difficult to justify major capital improvement programs when the financial resources can be used elsewhere with a better cost to benefit ratio.

NATURE OF DETERIORATION

Many years of exposure to a harsh northern US climate has taken its toll on many 60 to 80-year-old stadiums, such that their usable service life has been achieved. Reviewed herein are case studies of three football stadia – two collegiate and one professional facility. All three facilities have been investigated to evaluate the nature of deterioration, and two of the facilities have undergone substantial rehabilitation. On the remaining facility, rehabilitation is awaiting funding.

Structural rehabilitation of these facilities is necessitated for two primary reasons: deterioration of the concrete due to exposure to an exterior environment, and structural design/capacity considered insufficient when viewed under current building code loadings. Building code life-safety deficiencies usually also exist because of minimum aisle widths, egress exits, railing heights, and disabled spectator access and seating.

Concrete deterioration
Concrete material deterioration has been observed due to several mechanisms. These mechanisms are briefly reviewed below.

Freeze-thaw deterioration
This concrete deterioration mechanism occurs when moisture becomes trapped within a non-air-entrained concrete and the concrete containing this moisture is subjected to cycles of freezing and thawing. Concrete distress occurs when the water freezes and causes internal cracking of the concrete, usually at the exposed surface (called scaling). Once the concrete is cracked, additional water is able to penetrate and, upon freezing, causes additional damage. Deterioration of the concrete as the number of freezing and thawing cycles increase can be progressive with time.

Air-entrained concrete contains an air-void system purposely introduced with chemical admixtures, which is specifically intended to protect concrete from cyclic freezing damage. However, this admixture technology was not commercially available until the 1940s. Consequently, the concrete in these older stands is very susceptible to freeze-thaw damage when the concrete becomes saturated.

Consolidation
Oftentimes the concrete in these vintage stadia was site-batched and mixed to a stiff consistency (little or no slump). Consolidation of the concrete around reinforcing, at corners, or on form bottoms was not achieved because vibrators were not used. The resulting voids or honeycombed areas allow moisture to easily penetrate the concrete. Moisture ingress can easily initiate freeze-thaw damage to the concrete or reinforcement corrosion.

Carbonation
Older concrete structures exposed to the atmosphere are susceptible to concrete carbonation deterioration, which is a very slow chemical reaction taking years to manifest itself to measurable depths of any significance. The reaction starts at the exposed concrete surface and progresses inward. The reaction rate is influenced by the concrete's porosity, presence of cracks, and environmental conditions. Normally, a carbonation depth of about 10 to 15 mm (½ in.) in 50 years would be expected.

The reaction occurs when normal carbon dioxide (CO_2) in the area reacts with the concrete. The reaction lowers the concrete pH, which can reduce the protective nature of the concrete cover for the reinforcing steel. If the reinforcing bars have a shallow clear cover located in the carbonated zone, the bars can easily corrode when moisture and oxygen penetrate to the reinforcing bar level.

Chloride-induced corrosion
Concrete normally provides a high degree of corrosion protection to embedded reinforcing steel because of the concrete's high pH. The protective nature of the surrounding concrete's pH can be reduced due to the ingress of chloride ions to the steel surface. If moisture and oxygen are available to the reinforcing steel, the steel will corrode or rust. The corrosion mechanism by-products occupy a much greater volume in the concrete, which leads to delaminations and spalling.

Aggregate deterioration

Throughout the United States, aggregates exist that are not ideal for use in concrete. The aggregates themselves are subject to deterioration through alkali-silica reactions or freeze-thaw cycles, also known as D-cracking. An alkali-silica reaction in concrete occurs when the alkalis in the cement chemically react with available silica in specific aggregate types when moisture is present. The reaction product is expansive and continued reaction product productions eventually causes local concrete failure in the form of random or pattern cracking. D-cracking results when unsound aggregate becomes saturated, undergoes several freeze-thaw excursions, and deteriorates.

Design characteristics

Original design features causing recurring problems include those listed below.

Understrength

Original designs from the 1920s and 1930s were subject to few code design provisions. Modern day codes are quite prescriptive on loading requirements and the member capacity under ultimate load conditions. We have generally found the structural design adequate when viewed under current procedures. However, problems occur at localized members, connections, changed use and loading, or deteriorated locations. Strengthening is performed on an individual basis depending on the condition encountered.

Poor drainage

Early stadium designs intended all water to drain down the tread and risers toward the field. Collection inlets in the first row are often undersized, poorly maintained, clogged, or completely inoperable. Standing drain water is often observed in Row 1, which has lead to considerable deterioration.

Details

Certain details throughout the tread and riser system interrupt the free flow of drain water. Water entrapment is common, which can cause eventual deterioration. Water management around vomitories can be especially poor where water drains away from the field into the area below the stands. Freezing of water in this high traffic area is troublesome and dangerous.

Expansion joints

Modern expansion joint seals were non-existent when these older stadia were constructed. Folded sheet metal covers (galvanized steel or copper) embedded into the concrete often deteriorated after 20 to 30 years. These original details were also difficult to construct in a watertight manner. Usually this seal system was replaced with caulk that lasted one to two years. Yearly maintenance of the stadium consists of digging out some of the old caulk and gunning in more material. The new material usually has little flexibility, never bonds well to the dirty substrate, and is placed in a profile that will not accommodate the seasonal thermal movement.

Expansion joint location is another issue. When viewed from a sealing standpoint, some old joint locations are poor and are impossible to seal, even today. Examples of poor design or location include a conventional tread and riser expansion joint running through the center of successive vomitories, or a horizontally-oriented joint designed to accommodate appreciable movement in both principal horizontal directions.

STADIA REHABILITATION CASE STUDIES

Soldier Field – Chicago, Illinois

Located on Chicago's lakefront and opened in 1924, this reinforced concrete structure is one of the city's most architecturally historic buildings, featuring classic Roman colonnades above the main stands. The tread-and-riser system of the stands had deteriorated due to reinforcement corrosion and cyclic freezing and thawing of the non air-entrained concrete. Also, ground water fluctuations accelerated the deterioration of the highly permeable concrete at the base of the columns at or below existing grade. In the early 1980s, a rehabilitation program was undertaken that included several new amenities – new seats, scoreboards, locker rooms, natural turf, etc. The crux of the rehabilitation effort was strengthening and repairing of the stadium's structural system.

In many stand areas, concrete deterioration was quite advanced such that the traditional patching-and-waterproofing approach was impractical. Instead, a completely new tread-and-riser system was constructed using the old stands as a stay-in-place form. A 7.5-cm (3-in.) overlay was placed on the existing treads and new 15.0-cm (6-in.) wide risers were cast in front of the existing. The new system was designed to support its own weight, the weight of the existing stands, and the expected live load, a total load of over 14.4 kPa (300 psf).

Fig. 1. Load testing a tread and riser structural overlay at Soldier Field [1].

The added self-weight consequently necessitated strengthening of the existing concrete frames. Supplemental flexural and shear reinforcement was installed, and latex-modified concrete (LMC) was selected for the structural overlay, because it is less permeable to water than normal concrete. LMC also provided outstanding strength and durability, while bonding tenaciously to the existing concrete surface. This bonding aspect was important to provide composite action between the overlay and underlying girders, and to assure the old deck would be 'hung' from the new overlay.

To satisfy questions about the repair scheme performance and ability to achieve composite behavior, a trial installation and load test was conducted in an abandoned stadium section, as illustrated in Fig. 1. The trial installation carried a superimposed load of about 9.0 kPa (190 psf) without any signs of distress. Tensile bond tests conducted after the load test indicated very good bond between the overlay and existing stands. In spite of only minimal surface preparation, tensile failures generally occurred in the original concrete below the bond line [1].

After 15 years of service in this harsh northern climate, this repair scheme is still performing exceptionally.

Memorial Stadium – Lincoln, Nebraska
Located on the University of Nebraska campus in Lincoln, this classic war memorial stadium was constructed in 1923. The original stands are two separate, mirror-image grandstand structures located along each side-line. The cast-in-place, reinforced concrete structures, enclosed on the underside, had undergone two major rehabilitations in 1935 and 1945. Other patching programs were completed in 1955, 1986 and 1992. Both rehabilitation programs extensively used a dense shotcrete to cover nearly every exterior concrete surface – treads, risers, walls, columns, parapets, etc. However these overlays were not crack free and water infiltration through the original honeycombed concrete caused localized reinforcement corrosion. Freeze-thaw damage to the concrete and isolated pockets of alkali-silica reaction distress also occurred. These deterioration mechanisms caused the previous rehabilitation and repair work, and are partly responsible for the current work at this facility.

Traditional, modern concrete repair techniques have been employed, primarily to repair debonding and delamination of the 2.5 to 5 cm (1 to 2 in.) shotcrete overlay on the treads. The stands also contained a poor expansion joint blockout profile,
made worse throughout the years by poorly finished shotcrete edges and successive caulk installations. As shown in Fig. 2, the entire joint blockout was rebuilt to provide a more serviceable compression joint seal. A secondary seal system was also used to protect finished spaces below the stands.

Poor drainage management and clogged inlets resulted in continual water presence and saturated concrete in the first five rows of the stands. The highly advanced deterioration of the thin concrete tread made traditional patching unfeasible and uneconomical. A new structural overlay was installed, similar to the Soldier Field repair technique. The existing stands were used as stay-in-place forms, and new beams were formed and reinforced on the risers. A typical concrete placement operation is depicted in Fig. 3. The entire rehabilitated stands were coated with an elastomeric, urethane pedestrian membrane after concrete repair.

**Fig. 2. Rebuilding an expansion joint to provide
a suitable blockout profile for sealing.**

**Fig. 3. Casting a structural overlay using the
existing tread and riser system as forms.**

Ross-Ade Stadium – West Lafayette, Indiana

Purdue University's football stadium in West Lafayette, Indiana is a horseshoe-shaped facility of hybrid construction. The lower part of the bowl is cast-in-place, reinforced concrete consisting primarily of a slab-on-grade built into a low hillside. The higher extension of the stands is an open, riveted steel framework with steel plate tread and risers. The steel portion is well maintained and painted regularly. The concrete slab-on-grade has various degrees of deterioration present and has not received regular maintenance.

An interesting condition exists at this stadium; a urethane waterproof membrane was placed on the slab-on-grade concrete, presumably to protect the non air-entrained concrete, enhance the aesthetics, and provide slip-resistance. However, the membrane on the concrete throughout the stadium was very worn and exhibited numerous breaches. Control joints between various sections of the slab-on-grade were poorly constructed and sealed. Moisture detected beneath the concrete slab indicated the sub-base material was not well draining, and the drainage system was ineffective in diverting water away from the concrete treads and risers. Concrete core samples further revealed the concrete resting on grade had experienced long-term exposure to moisture due to the ground water.

The deterioration mechanisms were primarily attributed to moisture infiltration, a poor air void system in the concrete, and unsound aggregates. Unfortunately, the exposed surface membrane prevented evaporation of entrapped moisture, thereby allowing the concrete to remain continually moist or saturated. In a saturated state, this concrete was highly susceptible to freeze-thaw deterioration. Freeze-thaw deterioration in the concrete was accelerated by a small percentage of aggregates having poor freeze-thaw durability coupled with inadequate concrete air entrainment.

Rehabilitation of this stadium has not been initiated yet, as funding procurement is underway. Repair recommendations include removal and replacement of the severely deteriorated concrete tread and riser slab-on-grade sections. A new drainage system and free draining sub-base has been recommended.

SUMMARY

Rehabilitation of existing stadia is a feasible and economical means of renovating a facility to extend its service life. Fiscal constraints and desirable on-campus locations of existing facilities make rehabilitation more appealing. By focusing on some recurring deterioration mechanisms and their ability to be repaired, the authors believe new stadium designs can correct undesirable existing features.

Repair can prolong the design service life of older facilities. Plans for upgrading existing stadium facilities need to consider the structural system's condition along with the new amenities. Spending large sums of money on aesthetics and fan amenities while disregarding water leakage, deterioration, or areas of structural capacity concerns appears to be a short-sided investment. Furthermore, a defined and concentrated yearly maintenance program can hopefully address problems before they manifest themselves into conditions requiring large-scale rehabilitation projects.

REFERENCES

1. Klein, G. J., New Life for an Old Soldier: Rehabilitation of Chicago's Soldier Field, *Evaluation and Rehabilitation of Concrete Structures and Innovations in Design*, Proceedings ACI International Conference Hong Kong, 1991 (ACI SP-128, Volume I) Ed.: V. M. Malhotra, American Concrete Institute, Detroit, pp. 473–83.

5 ESTABLISHING BASIC FACILITIES IN PREVIOUSLY DISADVANTAGED AREAS OF SOUTH AFRICA

E. P. J. VAN VUUREN

BKS Consulting Engineers, Pretoria, and Rand Afrikaans University, South Africa

SUMMARY: This paper provides an overview of the development and establishment of basic sports and community facilities in rural South Africa. Identification of needs, community involvement, and design guidelines for effective planning are discussed.
Keywords: Basic sports facilities, rural communities, indoor sports, outdoor sports.

INTRODUCTION

Apartheid resulted in a sharp contrast in South Africa in standard and number of "black" and "white" facilities. Now the post-apartheid era faces a dilemma. Re-acceptance to international sport on the one hand demands the development of facilities that conform with international standards. Burgeoning urbanization and associated social upliftment programmes on the other hand call for the establishment of grassroots facilities – both at a time when funds are exceptionally scarce and other priorities clamour for attention.

The outlook is bleak. The tremendous backlog in housing and infrastructure in marginalized communities must be addressed. However, large spectator facilities also need to be developed, while existing facilities must be adequately maintained and managed to generate revenue and to cater for international sport in South Africa. A further complication exists with respect to financing the upkeep of grassroots facilities once they have been established.

South African sport can soar only on the wings of a fully integrated, community orientated, and financially viable strategy for the development of cost-effective, low-cost, functional facilities. Planning should address the individual aspects associated with the design, construction and long-term management of sports and community facilities in the South African context both systematically and comprehensively.

South Africa is desperately in need of facilities – in some cases very basic facilities. In squatter camps and informal settlements throughout the country, the authorities barely manage to provide the basic infrastructure for services such as water and sewerage reticulation. Recreational facilities in these areas rank low on the list of priorities.

Even in established black communities, very little exists in the line of sports and recreational amenities. Sprawling cities, such as Soweto with between one- and two-million residents, or Alexandra, are lucky to boast a solitary soccer stadium. Yet soccer is the most popular sport among black South Africans. Orlando stadium in Soweto can safely accommodate 30,000 spectators, but popular events may attract crowds of up to 60,000.

Training and coaching facilities are also virtually non-existent in these areas, with

Stadia, Arenas and Grandstands, edited by P.D. Thompson, J.J.A. Tolloczko and J.N. Clarke.
Published in 1998 by E & FN Spon, 11 New Fetter Lane, London EC4P 4EE, UK.   ISBN: 0 419 24040 3

ungrassed, levelled surfaces generally being used for this purpose. South Africa's participation in the Soccer World Cup finals makes the need for appropriate facilities to meet the expected surge in the popularity of soccer even more apparent.

SOCIAL RESPONSIBILITY

Professional integrity is, or should be, synonymous with social responsibility and should, therefore, be demonstrated by every member of the project team. However, social responsibility cannot be exercised in a vacuum. Sound understanding of local demographics and social dynamics is essential to:

- ensure maximized utilization with optimized resources
- proactively stimulate improved conditions and amenities.

Appropriate recreational facilities promote not only the physical well-being and personal fulfilment of the individual, but also the overall sociological health of the community. Escalating crime and violence in South Africa call for more and better leisure facilities, particularly at grassroots and mid-range level. The weight of responsibility in this regard rests on the shoulders of those engaged in the planning, design and construction of these amenities. However, the community itself also bears responsibility for the development, management and care of its own facilities.

African culture venerates the spirit of universal truth, morality and human dignity – traditionally referred to as 'Munisa-Mvula' (Rainmaker). Social responsibility manifests itself in a community where Munisa-Mvula is dominant. Western and African perceptions differ in that:

- the West sees social responsibility as conformance with prescribed and generally accepted norms and values
- Africa sees it as virtue arising from the heart of the people in solidarity, with Munisa-Mvula dominant and acting as the collective social conscience.

The Western approach offers the benefit of uniform measurement of actions and standard decision-making procedures. The disadvantage is its tendency to preclude non-standard alternatives possibly better suited to individual circumstances. Openness to alternatives arising from the people is the obvious advantage of the African approach; but Munisa-Mvula is not always dominant, hence the difficulty experienced by Westerners in identifying authentic community representatives, further complicated by changing political and civic structures due to volatile historical circumstances.

Those thought to be key representatives of the people frequently display two characteristics dominant in the West, namely:

- performance and enterprise (Muzwimi)
- power and conflict (Nndwa).

While the former is greatly admired from an African perspective, the latter is seen as negative but necessary to a limited degree. The affinity of these two characteristics to the

Western approach makes them conspicuous within Westernized structures, while those displaying the spirit of Munisa-Mvula, who are genuinely representative of the collective community, are often inconspicuous, resulting in cross-cultural interaction with power-bases and opportunists not truly representative of the people.

A further factor related to social responsibility is the need for local authorities to recognize the development of sports and leisure facilities as a social concern on a par with clinics, libraries, waste management, etc, thus calling for a regular allocation percentage-wise of the annual local and/or metropolitan budget. The argument for sport and recreation to be treated as a social responsible is particularly cogent in view of the rising rate of crime worldwide and universal acceptance that increased provision of leisure facilities has a significant impact on the social health and well-being of the community.

COMMUNITY LIAISON

Although specialist expertise is essential, community acceptance and involvement are no less important, the two working in tandem to ensure success. Thus the needs and aspirations of the community must be translated into achievable economic and technological goals to ensure:

- compliance with technical and functional standards
- appropriate solutions
- minimized building maintenance and repair costs.

Capacity building is often required in previously marginalized areas for effective community participation to take place.

On the other hand, technical input by specialists in the field needs to be tempered by community perceptions and participation to ensure an acceptable facility fully utilized by residents who feel pride in and responsibility for its proper management and maintenance. Thus community liaison and involvement should precede preliminary planning to encourage the willing co-operation of the people; not as an afterthought once most of the decisions have already been taken and much of the planning completed.

If the development of leisure facilities is to succeed as a harbinger of social well-being, community liaison is vital to:

- facilitate community acceptance, minimize vandalism, and ensure optimum use
- encourage responsibility for and ownership of the facility
- promote training and capacity building as a feature of the development itself
- identify available skills in the community
- publicize employment and small business opportunities
- ensure activities popular with both the community and those deemed to be socially delinquent, thus performing an effective sociological function.

Democracy in the Western sense is representative, with majority decision-making; in the African sense, it is participative, with consensus decision-making. This has significant implications where inter-cultural communication and liaison must take place.

Difficulties experienced in identifying authentic representatives can result in initial community liaison taking place with a shifting body of people, causing:

- lack of continuity
- duplication of effort
- power play between the changing role-players
- wasted time and delayed progress.

The possibility of changing political structures and role players should be borne in mind, therefore, even to the point, perhaps, of having an alternative plan available for emergency use, or organizing a carefully structured briefing session concerning the history of the developmental process to date for new role players entering the negotiation process at an advanced stage.

The benefits of open-air mass rallies to build community support and enthusiasm for a project is frequently overlooked. Where the mindset of the people is locked into an idea that really ought to be overturned in favour of a better alternative, it is an essential vehicle in convincing the community of the increased benefits, and winning their support.

HOLISTIC MASTER PLANNING

Sports and leisure activities should be fully integrated into the general town planning and physical characteristics of the area as a whole. Appropriate master planning, taking into account all contributory factors, is essential. However, a master plan must be dynamic, allowing for growth, demographic changes, and the usual vicissitudes of human nature.

In preparing a master plan, both the macro- and micro-environment must be catered for. Costly mistakes in the form of inappropriate, unacceptable or duplicated facilities can and should be avoided by undertaking a detailed community profile study at the outset.

MASTER PLANNING CONSIDERATIONS

Proper master planning should serve not only the long-term aspirations of the community and proper planning, but should also result in a document to serve as a basis for fund-raising. Donor funding (government or otherwise) can be solicited by means of a well-planned document incorporating a needs assessment, community and other stakeholders' acceptance, financial needs, and maintenance and management goals.

Preparation of a master plan should take into account all contributory factors, namely:

- demographic profiles and population forecasts
- government and municipal legislation pertaining to land-use
- current and anticipated patterns of participation, community opinions and preferences
- location and nature of existing facilities
- state of the existing facilities
- related development and structure plans
- public transport and development nodes in town and regional planning
- cultural profile and community needs
- other community needs and priorities

- financial considerations
- phased planning of facilities
- training needs.

FUNCTIONAL CONSIDERATIONS

Multi-functionality is essential in view of the limited resources in South Africa and the complexity of the current needs. Versatile indoor complexes can satisfy leisure, health, training, and educational needs by doubling up as sports venues, health clinics, and career training centres for optimum utilization.

Security and public safety currently pose a major concern throughout the world. Such multi-functional centres are able to accommodate pay-out points for pensioners, post offices, clinics, and other high risk activities, especially in rural areas.

The nature and layout of the facility depends on its intended purpose. Some facilities cater for informal community participation, while others cater for club, regional or provincial sport, each with its individual functional requirements.

INTERACTIVE COMMUNITY ANALYSIS AND MARKET RESEARCH

Interactive community analysis is essential as a preliminary exercise to establish a demographic profile of the marketplace and to correct misperceptions in the community and among role players and decision-makers.

Market research is a dynamic process which is significantly influenced, inter alia, by:

- individuals and their personal characteristics
- group dynamics
- cultural norms and values.

It is essential, therefore, to have input from experts au fait with sociological processes and categories of thought to ensure:

- questionnaires designed to transcend cultural and other limitations
- full understanding of each question by all respondents
- comprehension by the interviewer of the responses given
- realistic and culturally correct analysis and interpretation of the results
- quantification and presentation of facts for meaningful application.

Objectives of the market survey could include:

- a detailed description of the demographics and socio-economic profile of householders
- a broad overview of residents' sporting and shopping needs
- detailed analysis of residents' sports behaviour
- residents' views on proposed sites
- an indication of the development potential for various sports facilities.

Householders are asked standardized questions to yield a quantifiable response and to deal uniformly with all pertinent sociological and demographic factors. Face-to-face interviews, however, also allow for spontaneous responses and direct contact with the community.

The National Sports Council of South Africa is at present compiling a demographic map of South Africa, assessing and prioritising needs for all nine regions of the country. These documents are used to allocate funds and to make recommendations to prospective financial donors for the establishment of facilities.

FUNDING AND PLANNING OF FACILITIES

The Sports Trust is a privately funded, non-profit organization with trustees representing several large private companies, and the Department of Sport and Culture, National Sports Council, and National Olympic Committee of South Africa.

The Trust was established to fund basic facilities in disadvantaged areas of South Africa. It receives applications from communities, which are evaluated and prioritized by a committee which then allocates funds. The budget of the Trust is currently totally inadequate for the tremendous task that confronts South African sport. A dire need exists for more funds and the international community is urged to consider contributing in an effort to alleviate the situation. The Sports Trust provides an effective, secure avenue for distribution.

MULTI-SPORT VENUES

Once the demographic profile of a township or region has been completed, the local community is consulted to establish its priorities.

A master plan for the region follows and facilities are again prioritized. At this stage the community must apply for funds. A master plan and cost-estimate would normally be required to support the fund application.

INDOOR FACILITIES (See Fig. 1)

Indoor facilities should cater for the needs of rural communities, as they often constitute the only focal point or community centre serving large communities. They should be designed as sport centres catering for training, community participation, club and often regional competitions, cultural centres for concerts and religious gatherings, and community training centres, as well as for political or local government activities. Planning should attempt to cater for all the above needs at least in the long-term.

Typical example: The Sports Trust, typical indoor centre (see Figs 2 and 3)
The Sports Trust embarked on a visionary plan to provide basic facilities for the rural areas of South Africa. A basic multi-purpose hall was the result.

The typical indoor facility caters for sport, community activities, and often training, as well as cultural activities such as concerts and religious services, and becomes a focal point for the community. Design is the result of many years of experience in the master planning,

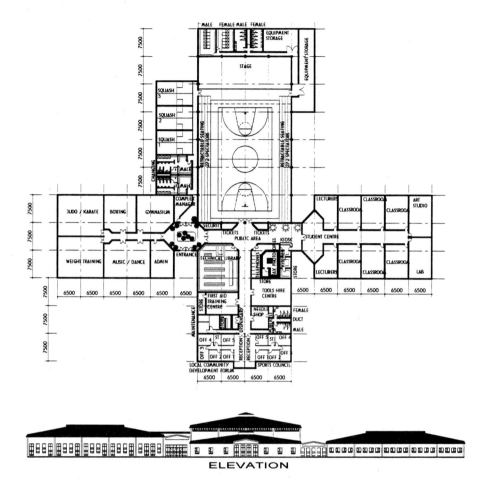

Fig. 1. Layout for multi-purpose indoor facility.

planning, design and construction management of such facilities by the author and his team of professionals.

Layout uses a basketball court as the basic size criterion. Limited permanent seating can be accommodated with the remaining space allocated to facilities for both spectators and competitors, such as change rooms, an office, security room, toilets, and a small kitchen.

The stadium is designed to minimize maintenance and to be as vandalproof as possible, calling for brick walls to roof level, with translucent sheeting to allow for natural lighting in the structure.

Permanently fixed louvers allow fresh air to enter the stadium at floor level and louvers at the apex allow the air to exit the stadium to maximize air flow. Subsequently, atmospheric measurements have been taken that have proven the method to be successful even in warm climates such as that of South Africa.

Local communities are typically consulted and resident skills identified. Local entrepreneurs then become responsible for the basic concrete structures, as well as the brick walls.

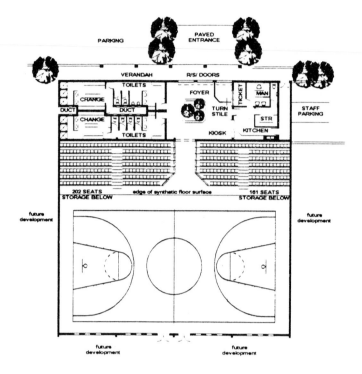

Fig. 2. Standard layout for indoor community hall.

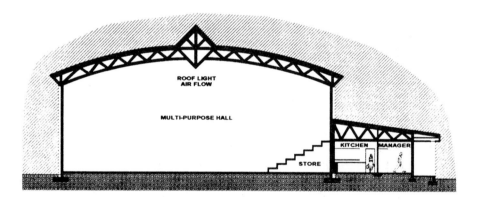

Fig. 3. Section through indoor community hall.

The Saulsville Arena (see Fig. 4)

Attridgeville has a population of 150,000 and is situated about 15 km to the west of the Pretoria city centre. The area is seriously lacking in basic sports and recreational facilities.

Demographic studies indicated not only a need for sports and recreational facilities, but also basic community needs, including a library, post office, community centre, training facilities and other facilities such as a video hire shop, clinic, post office and tool-hiring centre.

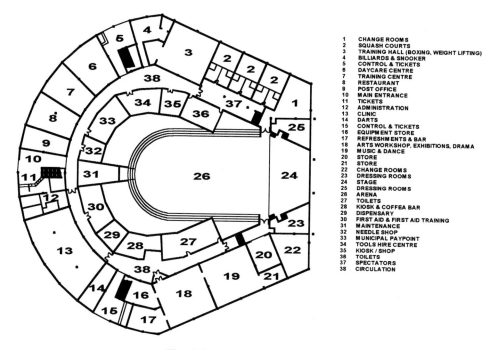

1	CHANGE ROOMS
2	SQUASH COURTS
3	TRAINING HALL (BOXING, WEIGHT LIFTING)
4	BILLIARDS & SNOOKER
5	CONTROL & TICKETS
6	DAYCARE CENTRE
7	TRAINING CENTRE
8	RESTAURANT
9	POST OFFICE
10	MAIN ENTRANCE
11	TICKETS
12	ADMINISTRATION
13	CLINIC
14	DARTS
15	CONTROL & TICKETS
16	EQUIPMENT STORE
17	REFRESHMENTS & BAR
18	ARTS WORKSHOP, EXHIBITIONS, DRAMA
19	MUSIC & DANCE
20	STORE
21	STORE
22	CHANGE ROOMS
23	DRESSING ROOMS
24	STAGE
25	DRESSING ROOMS
26	ARENA
27	TOILETS
28	KIOSK & COFFEA BAR
29	DISPENSARY
30	FIRST AID & FIRST AID TRAINING
31	MAINTENANCE
32	NEEDLE SHOP
33	MUNICIPAL PAYPOINT
34	TOOLS HIRE CENTRE
35	KIOSK / SHOP
36	TOILETS
37	SPECTATORS
38	CIRCULATION

Fig. 4. Saulsville Arena layout.

An existing soccer stadium for 50,000 spectators had been closed for matches due to the unsafe steel structure. The Saulsville arena, as it is known, was the only venue that could be used for cultural activities. It was, however, dilapidated and unfit as a venue for social upliftment.

Plans were prepared to convert this venue to an indoor stadium fit for multi-purpose utilization. The floor of the arena was raised to increase the plan area to accommodate a full basketball court. This would afford the opportunity to provide floor markings for basketball, volleyball, badminton and tennis. To accommodate the larger floor space the height of the floor was raised by about 1 m, sacrificing three rows of spectator seating. A synthetic floor is planned for the final design.

The existing roof will be removed and replaced with a steel roof to cover the entire arena, creating an indoor arena. Seating was increased to 4,000. Commercial activities were considered, but finally rejected as the site was not viable for that type of activity.

The City Council of Pretoria is at present in the process of raising the estimated £1.0 million to complete the first phase.

OUTDOOR FACILITIES (See Fig. 5)

The popularity of sport among rural South Africans often still remains divided by racial preferences, although this is changing. The popular outdoor sports codes are soccer, cricket, basketball, volleyball, tennis, bowling, and rugby. The United Cricket Board, for instance, has made a tremendous contribution by marketing cricket to all sectors of the South African population. Dr Ali Bacher, through his enthusiasm and vision, has established a number of

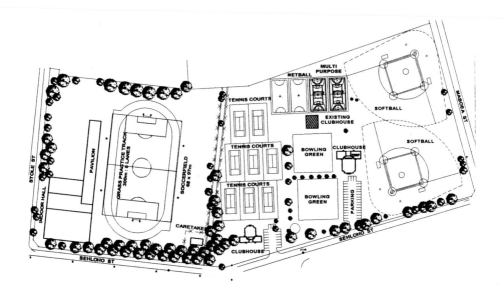

Fig. 5. Typical outdoor layout.

basic facilities and has organized training sessions in even the most remote areas.

The master plan should include all the sporting needs for a particular community, as well as training, management and maintenance needs. Basic facilities can then be established according to a structured plan as funds become available. Basic multi-purpose, low-maintenance courts and pitches are necessary.

Soccer fields are grassed, but sub-surface drainage is often too expensive and only surface drainage can be afforded. Most soccer fields are over-populated due to the popularity of the sport, calling for intensive maintenance.

Synthetic surfacing would provide a solution to the higher population usage of the fields, but costs are too high to justify this medium. Kikuju grass, an exceptionally sturdy grass, is employed on a sand layer for drainage purposes.

Most rural and under-privileged South Africans travel vast distances to and from work, generally leaving home in the early morning and returning late in the day, with the result that sport is practised at night, calling for lights at sports venues.

Typical example: Klipspruit Soccer Centre

The South African Soccer Development Centre at Klipspruit, Soweto, was established with a grant from Foodcorp Limited.

The centre covers a surface of 4 ha and comprises four soccer fields (105 × 68 m). One of the fields serves as a main field seating 300 spectators on concrete. Next to the field is a fully equipped clubhouse with conference and lecture facilities, change rooms, kiosk, storerooms, a first-aid room, and offices.

CONCLUSION

Following the learning process involved in this and several other cases, the author is convinced that a synthesis must be achieved between the Western and African cultural approaches to ensure effective development of urgently needed sports and leisure facilities appropriate to specific community needs and acceptable to the people. The benefits of each approach need to be harnessed and the disadvantages overcome, with mutual respect and understanding throughout the consultative process.

Ubuntu implies finding and affirming one's identity in relation to others within our rainbow nation. It implies nation-building by a people invigorated by a collective vision for the future that is both technically and economically feasible.

ACKNOWLEDGEMENTS

The author acknowledges reference to the book by Lovemore Mbigi and Jenny Maree entitled 'Ubuntu: The Spirit of African Transformation Management', published by Knowledge Resources (Pty) Ltd in Randburg (1995). Shona terms have been adapted to suit the local context.

REFERENCES

Articles
1. Van Vuuren, E. Building our Nation............. From the Grass Roots, *Official Yearbook of South Africa Sport 1996*. National Sports Council.
2. Van Vuuren, E. Master planning for a Sporting Chance. *MIE Municipal Engineer*, January 1996.
3. Van Vuuren, E. Wings of Hope. *Panstadia International*, Summer 1993.
4. Van Vuuren E. Spoelstra Tjibbe. Safety in Numbers. *Parks and Grounds*, No. 76, 1993.
5. Van Vuuren, E. Planning Effectively for Sporting Facilities, *Parks and Grounds*, No. 70, 1992.
6. Van Vuuren, E. Home Ludens. *Building*, No. 57, October 1966.
   Louw J P, Van Vuuren E. The design of indoor sports arenas. *Parks and Grounds*.
   Bruinette K E, Van Vuuren E P J. Design of Sport Stadia. *The Civil Engineer in South Africa*, February 1978. (Afrikaans).

Symposia
1. Van Vuuren, E. Community Participation: The Balancing Act. *Conference for Parks and Recreation*, Bloemfontein, 1996.
2. Van Vuuren, E. *Symposium to promote awareness of Sport and Recreational Facilities in Mpumalanga.* Barberton 1996.
3. Van Vuuren, E. The Criteria to be considered for the Design of Indoor Sport / Cultural Centres. *Convention of Facilities for Sport and Recreation*, Johannesburg, January 1996.
4. Bruinette K E, Van Vuuren E P J. Design Criteria for Sport Stadia (Afrikaans). *Proceedings of the Sixth Quinquennial Convention of the SAICE*, Durban. 1978.

PART TWO

RISK MANAGEMENT AND SAFETY

6 THE 'GREEN GUIDE' AND ITS SIGNIFICANCE

J. R. K. DE QUIDT and S. THORBURN
Football Licensing Authority, London, UK

SUMMARY: This paper examines the need for and evolution of the Green Guide, its underlying philosophy and its key themes. It shows how to use the Guide and discusses the impact of the new edition.
Keywords: Capacity, circulation, disaster, disorder, guidance, responsibility, safety, spectators, sports grounds.

INTRODUCTION

In 1997 the Government published the Fourth Edition of the *Guide to Safety at Sports Grounds*, commonly known as the 'Green Guide' [1]. This was the product of a comprehensive review of available research and good practice. It was thus the first version not to be issued in haste in response to problems or in the aftermath of a disaster.

In this paper we explain how the Guide first appeared and why it was needed. We chart its evolution into an authoritative detailed guidance document. We identify its underlying philosophy and main themes and explain how it should be used.

THE EVOLUTION OF THE GREEN GUIDE

Disaster and disorder
Government guidance on sports ground safety has to be seen in the context of the major disasters at British football grounds or involving British supporters. The most serious such incidents since 1945 are listed in Table 1.

The 1970s and 1980s also witnessed significant disorder in and around many football grounds. This combination of safety and public order problems led to the Government taking a progressively greater regulatory role, culminating in the Football Spectators Act 1989, the requirement that grounds in the English Premier and First Divisions and the Scottish First Division become all-seated, and the creation of the Football Licensing Authority.

Stadia, Arenas and Grandstands, edited by P.D. Thompson, J.J.A. Tolloczko and J.N. Clarke.
Published in 1998 by E & FN Spon, 11 New Fetter Lane, London EC4P 4EE, UK. ISBN: 0 419 24040 3

Table 1. Deaths and injuries in major accidents at British football stadia since 1945

Year	Stadium	Cause	Deaths	Injuries
1946	Bolton	Crushing	33	400
1957	Shawfield, Glasgow	Barrier collapse	1	50
1961	Ibrox Park, Glasgow	Crush on staircase	2	50
1961	Huddersfield	NK	1	?
1962	Oldham	Barrier collapse		15+
1963	Arsenal	Crushing		100+
1964	Port Vale	Fall/Crushing	1	2+
1964	Sunderland	Crushing		80+
1966	Anfield, Liverpool	Crushing		200
1967	Leeds	Crushing		32+
1967	Ibrox Park, Glasgow	Crush on staircase		8
1968	Dunfermline	Barriers collapse	1	49
1969	Ibrox Park, Glasgow	Crush on staircase		24
1971	Ibrox Park, Glasgow	Crush on staircase	66	145
1971	Oxford	Wall collapse		25
1971	Stoke	Crushing		46
1972	Arsenal	Crushing		42
1972	Wolverhampton	Barrier collapse		80
1975	Lincoln	Wall collapse		5
1978	Leyton Orient	Barrier/wall collapse		30
1980	Middlesbrough	Gate pillar collapse	2	?
1981	Hillsborough, Sheffield	Crushing		38
1984	Walsall	Wall collapse		20
1985	St Andrews, Birmingham	Wall collapse	1	20
1985	Valley Parade, Bradford	Fire	56	Hundreds
1885	Luton	Disorder		47
1987	Easter Road, Edinburgh	Crushing		150
1989	Hillsborough, Sheffield	Crushing	96	400+
1989	Middlesbrough	Crushing		19
1993	Cardiff Arms Park	Distress rocket	1	0

Ibrox - the Wheatley Report

On 2 January 1971 66 spectators died at Ibrox Park, Glasgow in crushing on a stairway. This was the fourth major incident, and the second involving fatalities, on this stairway in ten years. Fig. 1 shows the damage to the barriers on the stairway caused by the crowd pressures which led to the disaster.

The Government set up an Inquiry under Lord Wheatley. His terms of reference were to make an independent appraisal of the effectiveness of existing arrangements for crowd safety at sports grounds and of possible improvements within the law as it then stood; and to consider how the law needed to be altered.

Although Lord Wheatley concentrated on safety at football grounds, he realised that rugby grounds had potentially the same problems.

Fig. 1. Damage to barriers at Ibrox Park Stadium, January 1971.

Absence of guidance

It quickly became clear that there was a complete lack of guidance to both the management of football clubs and their technical advisers on how to alter or improve their grounds. They had to rely on the advice of consultants and contractors and could not be certain that they were spending their money wisely. Identical problems were being addressed in very different ways between one ground and another.

Lord Wheatley formed a Technical Support Group to examine any advice submitted to the Inquiry and to conduct research on whether and if so what guidance could be given. This produced recommended guidelines, based upon the latest knowledge available, to assist both clubs and local authorities.

First edition of the Green Guide

The First Edition of the Green Guide duly appeared in 1973 [2]. It set out areas of concern and made recommendations, mainly relating to football. It was, however, a simple document. Stadium designers had still to supplement it from their own experience.

The Government decided that advice alone was not enough. Accordingly, the Safety of Sports Grounds Act 1975 [3] established a framework of safety certification by local authorities for larger sports grounds (broadly those with a capacity of at least 10,000). With a few modifications, this is still in use today.

Safety in the round

Lord Wheatley formulated two important conclusions which are still highly relevant.

Danger can arise not merely from the structure or configuration of the ground but also from the unpredictable nature of human behaviour in unexpected situations. This cannot be governed by rules. It requires a management system able both to cater for all normal conditions and to respond to reasonably foreseeable circumstances.

The problem of crowd safety cannot be solved simply by looking at part of the ground in isolation. Each part has to be looked at, then the inter-relation between parts, and then the overall picture. Deficiencies in certain aspects may be off-set in whole or in part by other aspects.

Moreover, while little or no danger may arise when the normal attendance at a ground is only a small proportion of its total capacity, dangers not normally present may be brought into existence if there is a capacity or near capacity crowd.

Lord Wheatley also particularly highlighted the dangers of crowd pressure in standing areas and of fire in seated stands. Tragically, his words were not sufficiently heeded.

Football hooliganism

Lord Wheatley's vision was overshadowed by the growth of 'football hooliganism' in the mid 1970s. In 1976 the Green Guide was supplemented by further guidance for football on matters such as segregation, alcohol and all-ticket matches. The focus of attention moved increasingly from safety to public order. It would take two more major disasters, at Bradford and Hillsborough, before the conclusions enunciated by Lord Wheatley were fully taken on board.

Bradford - the Popplewell Report

On 11 May 1985, 56 people died in a fire at the Valley Parade ground in Bradford. The Government set up a public inquiry under Mr Justice Popplewell.

His interim report [4] graphically described how and why the fire started and why there were so many casualties. It found that the stand was a wooden structure, with a void under the seats, in which debris could and did collect and that the available exits were insufficient to enable spectators safely to escape the devastating effects of the rapidly spreading fire.

The report noted that many people believed erroneously that the Green Guide applied only to the largest sports grounds. "Had the Green Guide been complied with this tragedy would not have occurred."

The Second Edition of the Green Guide was published in 1986 following the final Popplewell report [5]. Not surprisingly, this laid particular emphasis on fire safety. It also included new material on temporary stands and marquees.

Again this guidance was supplemented by legislation. The Fire Safety and Safety of Places of Sport Act 1987 [6] introduced a system of safety certificates for all stands with a capacity of at least 500 at sports grounds not covered by the 1975 legislation.

Hillsborough - the Taylor report

Yet even now, a culture of safety failed to take hold. On 15 April 1989 English football reached its nadir at Hillsborough in Sheffield. Ninety five people were crushed to death during an FA Cup semi final. Fig. 2 shows a view of the terracing involved in the 1989 disaster.

The Hillsborough ground was regarded at the time as one of the best in the country. However, with hindsight, warning signals were missed. There had been a serious crushing incident at the ground in 1981. Moreover, disaster had been avoided at the 1988 semi-final between the same two teams by good fortune and not by good management.

Lord Taylor noted with dismay that his was the ninth official report [7] covering crowd safety and control at football grounds. Yet the lessons of past disasters and the subsequent

Fig.2. Leppings Lane standing terrace at the Hillsborough Stadium.

recommendations had not been recognised. Many of the deficiencies at Hillsborough were in areas that had been addressed by previous reports.

AFTER HILLSBOROUGH

Radical change
The Hillsborough Disaster and the Taylor Report [7] marked the turning point. Lord Taylor emphasised "that the years of patching up grounds, of having periodic disasters and narrowly avoiding many others by muddling through on a wing and a prayer must be over. A totally new approach across the whole field of football requires higher standards both in bricks and mortar and in human relationships."

Third Edition of the Green Guide
The Taylor Inquiry's Technical Working Party advocated a much higher degree of standardisation in respect of new work at sports grounds. It also advised that the Green Guide should form the basis on which the local authority issued a safety certificate.

Any departure from its guidance should be fully defined and approved in writing by the local authority.

The Third, greatly expanded, Edition of the Guide was published in 1990 [8] against this background. It placed a much greater emphasis on the responsibilities of management and on crowd control. In keeping with Taylor's vision of civilised surroundings open to all kinds of spectators the new edition also, for the first time, covered the subject of spectators with disabilities.

Football Licensing Authority
The Football Licensing Authority (FLA) was created by the Football Spectators Act 1989 [9]. Its two key functions are to ensure that all the grounds in the English Premier and First Divisions become all-seated (now all but complete) and to oversee safety certification by local authorities. The Green Guide is a key tool in both these areas.

The FLA has become the main source of advice and guidance to local authorities, clubs and many other bodies on safety at sports grounds. It has been able to bring about a much greater degree of consistency, above all in the calculation of safe capacities. The enormous improvements in both the structure, the facilities and the safety management of professional football grounds over the past few years have been well documented.

There has been a dramatic fall in the number of injuries to spectators. The first aiders are now mainly finding themselves treating the sick. For the first time football grounds are being managed proactively: those in charge are no longer merely reacting to events. The challenge is now to ensure similar improvements in other sports.

Need for a new edition

The Third Edition stood the test of time well. It was widely used and respected and proved a key element in the improvements of the early 1990s. However, like all post-disaster material, it suffered from having to be issued quickly. Inevitably some sections were ambiguous or unclear. It was also necessary to take account of current good practice and of the improvements in safety and ground management of recent years.

The Football Licensing Authority was therefore commissioned by the Government to co-ordinate a review and revision of the Guide. It set up a Working Group, including outside experts, to identify and eliminate ambiguities and contradictions, to spell out clearly what was previously only implicit, and to take full account of recent research and experience.

The Working Group was specifically instructed that it should resist any temptation to ratchet up standards. This requirement was observed. While the new edition is longer and more detailed than any of its predecessors, no sports ground complying fully with the recommendations in the Third Edition would have to be upgraded to meet those in the new edition.

THE FOURTH EDITION OF THE GUIDE

Underlying principles

Two fundamental principles, first enunciated by Lord Wheatley, underlie the Green Guide. The first is that safety cannot be achieved by means of regulations imposed upon the reluctant from outside. Instead those who are responsible must believe in it for themselves. Individuals and organisations must develop a safety culture.

The second principle is that safety is not merely to do with structures. It also concerns the management both of the structures and of the staff and spectators within and around them. This management must be proactive, identifying and resolving problems before they occur.

Responsibility for safety

The Guide emphasises that it is the ground management, normally the owner or lessee of the ground, who may not necessarily be the promoter of the event, who is responsible for spectator safety. In certain sports, most notably football during the 1980s, there has been some confusion over who was responsible for spectator safety. This left a vacuum which came to be filled by the police as the only organisation with the communications and manpower to cope. This vacuum still exists in certain sports.

The Guide is therefore addressed primarily to ground management and its professional advisers, and to the various regulatory authorities. It enables competent people to reach reasonable decisions having regard to local circumstances. Its purpose is to empower and not to constrain those who use it.

Achieving reasonable safety

The Guide recognises that absolute safety is a chimera. A sports ground is absolutely safe only when it is empty. The guidance is therefore directed at ensuring reasonable safety. One of the main tools used in determining what is reasonably safe is the risk assessment. This should be carried out not by the regulatory authority but by the person who is responsible for safety.

The Green Guide does not purport to answer every question. There will be occasions where an alternative approach may well be justified. In such cases, however, the onus is on those seeking to follow another way to demonstrate that this will provide an equal or greater level of safety. The Guide is a distillation of many years of research and experience in the design and management of sports grounds, from which nobody should depart lightly.

Safe capacities

The primary objective of the Green Guide is to provide guidance on the safe capacity of a sports ground, that is the number of spectators whom it can safely accommodate during an event. Contrary to what might be supposed, the safe capacity is not necessarily the same as the physical capacity, for example as measured by the number of seats.

Other factors which have to be taken into account include the capacities of the entrances, circulation areas and exits, the physical condition of the premises and the quality of the safety management. The Fourth Edition spells these out in detail for the first time.

For example a lower division football ground might physically be able to accommodate say 10,000 people, but its systems and personnel might only be capable of managing its normal attendance of 3,000. In such a case, the lower figure represents the safe capacity.

The management of safety

As Lord Wheatley observed, safety at sports grounds requires both good management and good design. As with the calculation of safe capacities, much of the emphasis must be on the active management of spectators. The moments of greatest danger at sports grounds, and those when most disasters have occurred are when spectators are entering, leaving or on the move. A moving crowd is both more difficult to manage and more vulnerable to crushing. The Guide therefore devotes substantial sections to crowd management and to the design of means of circulation and protective barriers.

New guidance

Early editions of the Green Guide provided guidance on measures intended to improve safety at existing grounds. The Guide has never been a design manual for new facilities. Nevertheless, during the preparation of the Fourth Edition it became clear that some guidance was needed on good practice in the design and management of new grounds or newly-built parts of grounds. This was therefore included in the Guide for the first time.

These sections are clearly identified so as to avoid the danger of local authorities seeking to apply unreasonable requirements to existing facilities. Where existing structures are safe but of a lower standard, for example standing terraces with poor views, this can be addressed by reducing the capacity using the formula set out in the Guide.

Seat row depths

Three particular examples of guidance directed to new building are worthy of mention. The first concerns seat row depths. Earlier editions of the Guide had identified minimum row depths of 610 mm, without specifying that this was in practice acceptable only for bench seating with no backs.

The Third Edition suggested depths of 760 mm for comfort but otherwise stuck with 610 mm. The new edition, while continuing to recommend 760 mm, accepts 700 mm as a reasonable minimum where the higher figure cannot be achieved.

Structural dynamics

The Working Group revising the Guide was specifically asked to examine the issue of robustness and structural dynamics which was causing concern in some quarters. This is a complex subject on which the various experts do not always agree.

The Working Group decided to recommend values for dynamic stiffness, based on practical experience, which may be adopted by designers of sports grounds seeking to achieve reasonable safety at reasonable cost. It did not lay down serviceability criteria since it was not possible to state with any certainty what values of accelerations which would be accepted by crowds deliberately exciting a structure. Further research is needed on this subject.

The guidance in the Fourth Edition represents the current conventional wisdom. Many stands have been built during the past 25 years on the basis of the standards recommended. No evidence was produced of any structural malperformance, though, in a few cases, a lack of dynamic stiffness had led to reports of serviceability problems. Accordingly the Guide emphasises that careful consideration should be given to dynamic effects before sports grounds are used for certain types of concert.

Spectators with disabilities

The Building Regulations prescribe the proportions of places that should be provided for spectators in wheelchairs in new sports grounds or stands. For undefined larger facilities an unspecified lower figure is permitted. Despite exhaustive enquiries, the Working Group was unable to ascertain the rationale for the particular prescribed figures.

With the agreement of the Government therefore, it included in the Fourth Edition detailed recommendations as to the provision of wheelchair spaces at sports grounds with a capacity above 10,000, based on current good practice.

IMPACT OF THE GUIDE

Football

The Fourth Edition of the Guide has been widely welcomed, in particular by football clubs, their advisers and local authorities. It is comprehensive, clear and easy to read and apply. By bringing together in a single document much material that was previously difficult to locate, it has greatly facilitated their task.

Paradoxically it may also signal the beginning of a reduction of the Government's direct involvement in the affairs of football clubs. By emphasising the positive role of ground management and the use of risk assessments, it puts the responsibility back where it belongs.

As a result of the work of the Football Licensing Authority, the capacities of all Premier and Football League grounds in England and Wales are already calculated in accordance with the procedures laid down in the Guide. Their structures and management also generally reflect the guidance in the Guide. While there is still much to do, a pattern has now been established. The main concern is to prevent bad habits returning.

Other sports

In other sports, however, the picture is more patchy. Early editions of the Green Guide focused very much on the safety of spectators at football grounds. The argument is still sometimes heard that certain safety requirements are unnecessary "because our spectators are different – we do not have a hooligan problem".

This view is fundamentally misconceived. It betrays a misunderstanding of the principles of safety and a confusion between safety and public order. While their specific circumstances may vary, the principles of safety are the same. The new edition emphasises that the Green Guide applies to all sports.

Overseas

The Guide is also beginning to have an impact overseas. It is being used by the Working Party producing European standards for stadia. It has also been used by architects for various major projects overseas where equivalent guidance is not available.

CONCLUSION

In his report of his Inquiry into the Hillsborough Stadium Disaster, Lord Taylor identified complacency as the main enemy of safety. The Green Guide by itself cannot overcome complacency. However, in the hands of those who are committed to safety, it is an invaluable tool.

REFERENCES

1. *Guide to Safety at Sports Grounds, Fourth Edition*. HMSO, London. 1997.
2. The Rt. Hon. Lord Wheatley. *Report of the Inquiry into Crowd Safety at Sports Grounds*. HMSO, London. 1972.
3. *Safety of Sports Grounds Act 1975*. Chapter 52. HMSO, London. 1975.
4. Mr Justice Popplewell. *Interim report of the Committee of Inquiry into Crowd Safety and Control at Sports Grounds*. HMSO, London, 1985.
5. Mr Justice Popplewell. *Final report of the Committee of Inquiry into Crowd Safety and Control at Sports Grounds*. HMSO, London, 1986.
6. *Fire Safety and Safety of Places of Sport Act 1987*. Chapter 27. HMSO, London. 1987.
7. The Rt. Hon. Lord Justice Taylor. *Final report of the Inquiry into the Hillsborough Stadium Disaster*. HMSO, London, 1990.
8. *Guide to Safety at Sports Grounds, Third Edition*. HMSO, London. 1990.
9. *Football Spectators Act 1989*. Chapter 37. HMSO, London, 1989.
10. *Construction (Design & Management) Regulations 1994*. CONDAM/CDM Regs. HMSO, London, 1994.

7 STRATEGIC RISK MANAGEMENT IN PUBLIC ASSEMBLY FACILITIES

S. J. FROSDICK
IWI Associates Ltd, London and University of Bradford Managment Centre, Bradford, UK

SUMMARY: Managing hazards and risks is an integral part of public assembly facilities (PAFs) management. This paper looks at PAFs as systems broken down into five zones: the event area, viewing accommodation, inside concourses, outside concourses and the neighbourhood beyond the venue. It draws out the different types of hazards perceived in each zone by each of four different groups of stakeholders. Owners and operators are concerned with threats to their revenue streams. Spectators wish to view and enjoy events staged in comfortable surroundings. Regulatory agencies seek to enforce safety and security rules and the community want their environment disrupted as little as possible. The paper suggests that PAFs managers need to recognise the legitimacy of these different points of view and concludes by offering a practical management approach which allows for greater richness, diversity and consensus in the analysis.

Keywords: Hazard, public assembly facilities management, risk, risk perception, strategic risk management.

INTRODUCTION

Stadia and arenas, increasingly referred to as public assembly facilities (PAFs) [1], are venues of extremes. On the one hand, they are the setting for some of the most exciting, enjoyable and often profitable events in the world. On the other, they have been the scene of many terrible disasters [2]. In 1966, 300 people died in a stadium riot in Lima in Peru. In 1992, 13 people were killed when a temporary stand collapsed early in a football cup semi-final in Corsica. In June 1996, 15 football fans died when a wall collapsed in a crowd stampede during a world cup match in Zambia. Not to mention either Heysel in Belgium or the catalogue of disasters involving the deaths of hundreds of British football supporters. The many well-documented near misses, for example the collapse of a stand during a Pink Floyd concert at London's Earl's Court arena, are further evidence of the potential for PAFs disasters.

Dealing with safety hazards and the risks to which they give rise must therefore be seen as an important part of PAFs management. Operators have a moral and legal duty of care towards participants, performers, spectators and staff alike. In some countries, they have a statutory duty to deal with risks to the health and safety of people who might be

Stadia, Arenas and Grandstands, edited by P.D. Thompson, J.J.A. Tolloczko and J.N. Clarke.
Published in 1998 by E & FN Spon, 11 New Fetter Lane, London EC4P 4EE, UK.   ISBN: 0 419 24040 3

affected by the operation of their facility. Following several high-profile legal cases, there is growing awareness of the civil and even criminal liability of PAFs management in the event of a disaster precipitated by negligent preparations [3, 4].

Thus the importance of risk assessment has been brought into ever sharper focus. Understandably, the emphasis has been on public safety hazards, and whilst these must be paramount, it can be shown that looking at them in isolation can create operational difficulties for PAFs management.

Different perceptions

The problem is that managers are faced with four competing demands. Commercial pressures require them to optimise the commercial viability of the venue and its events. Spectator demands for excitement and enjoyment require credible events staged in comfortable surroundings. Regulatory and other requirements for safety and security must also be met, whilst any negative effects which the venue and event may have on the outside world must be kept to a minimum.

Each of these areas contains sources of hazards and risks. My own research into strategic risk management in PAFs in Britain has shown how risk means different things to different people in different contexts [5]. Regulatory perceptions of risk as breaking safety rules are predominant, and a multitude of agencies – emergency services, local authorities, governing and enforcement bodies – are involved in safety and security management. PAFs owners and operators are more entrepreneurial and give priority to commercial risks such as access control, pirate merchandise, ticket touting, cash handling and ambush marketing. Spectator and local residents pressure groups are more concerned with quality and environmental risks. The vociferous minority of spectators demand the right to sit (or stand) where they choose to watch the event as they please without being commercially exploited or having their enjoyment intruded upon by petty-minded officialdom. Local residents voice concerns about the impact of noise, litter, traffic, vandalism and parking. The majority of spectators tend to shrug their shoulders with a fatalist acceptance of the various hazards they endure as a result of the commercial, regulatory, and behavioural excesses of all the others.

Operational conflicts

Successful PAFs management means striking an appropriate balance between these demands. Yet there are many examples of the operational problems which have arisen as a result of management failure to achieve this balance.

At a Newcastle United versus Sunderland football match in 1993, disorder broke out as a result of stewards and the police removing a Sunderland banner draped over a pitchside advertising board. The banner was preventing the sponsor's name from getting television exposure. Perceiving the hazard, the commercial manager had deployed the stewards without consulting anybody. The police had pitched in to help the stewards when they perceived a public order hazard. Two officers snatched the banner and a fight broke out. I was watching from the control room with the stadium safety officer: he was furious at the safety hazard created as members of the crowd suffered the results of the fight.

A second example comes from 1995 from a football ground in the north-west of England. Part-way through briefing the senior stewards, the safety officer was called away to speak urgently to the commercial manager. The latter told him there was a fire in a hospitality suite and requested his immediate attendance. The commercial manager was teasing – "I thought that would get you here quick". In fact he wanted the pitch covers to be moved from where they had been folded up because they were preventing the advertising hoardings being seen. In dealing with the commercial risk, the manager thought nothing of disrupting the essential briefing for the senior stewards, which had to be curtailed. What would any subsequent inquiry into a real fire have made of the disruption and the irresponsible lie?

These examples illustrate operational conflicts between safety and commercialism. Of course there are other conflicts, for example between safety and enjoyment, such as the indiscriminate banning at some venues of the flags and drums which add so much to the carnival atmosphere at sporting events. But the two examples from my own experience must suffice to make the point here.

A STRATEGIC AND SYSTEMS APPROACH

My argument, then, is that to balance these differing demands and perceptions, the PAFs manager needs to adopt a more strategic and holistic approach to hazard and risk assessment [6, 7]. What this means in practice is an acceptance that nobody is wrong either to perceive a particular issue as a hazard, or to evaluate a risk in a particular way. It is therefore important to ensure that a broad range of perspectives are adequately represented in any risk assessment exercise [8]. This is best accomplished in two ways. Firstly, any exercise should be undertaken by a group of people, rather than just one or two. Secondly, representatives of each of the four groups: commercial, regulatory, spectator and local resident, should be identified and invited to participate.

To illustrate the potential richness and diversity of this approach, I want to look at PAFs as systems, broken down into zones, such as is shown in Figure 1.

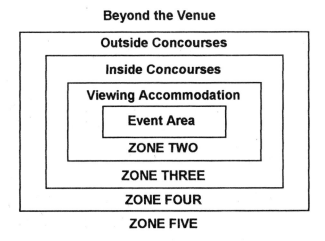

Figure 1. Public assembly facilities as a system of zones.

My site visits and analysis of briefing documents, contingency plans and training material has enabled me to catalogue, probably not exhaustively, the variety of hazards which have been perceived in each part of the system.

So let me begin by looking at the most commonly perceived hazards in the area where the event is held (the pitch, track, or court, etc.) and the perimeter between it and the viewing zone.

The event area and perimeter

The commercial perspective

Threats to the interests of advertisers and sponsors form the principal sources of event area hazards perceived from a commercial perspective. The two examples cited above show the commercial importance of ensuring that perimeter advertising is clearly visible to the television camera.

Since accredited sponsors will have paid substantial fees to be associated with the event, there is also a perceived need to prevent 'ambush marketing' by other brands. At the Portugal versus Turkey match in Nottingham during the recent European football championships, several banners advertising Portuguese products were brought into the stadium and displayed whenever play and thus the cameras went in their direction. Stewards had to be more active dealing with these banners than they did with the well-behaved crowd!

Television companies also pay handsomely for their access and are inevitably anxious to minimise the risks of high installation costs and poor broadcast quality in their choice of camera positions and cable runs around the event area perimeter. Conflicts arise when these choices create trip hazards or obstruct spectator sightlines.

The spectator perspective

Since their main purpose is to watch the event, any deficiencies in sightlines, in the physical event area and in the event itself provide sources of hazard to the enjoyment of the spectators. Restricted views arise from old PAFs designs, with roof props and even floodlight pylons around the perimeter of the event area. Unusually high perimeter hoardings, cage-type fences, inappropriately sited television cameras or excessive deployments of police or stewards around the perimeter represent further sources of hazards to sightlines.

Sports events may either become a farce or else be unplayable if surfaces, particularly grass, become too wet. The high jump section of the women's pentathlon competition at the Atlanta Olympics, where standing water was not properly cleared from the runway, adversely affecting the athletes' performances, provides an example. Enjoyment may also be threatened by a lack of credibility in the event itself. In boxing, a number of 'big fights' have ended in the first round because of the mismatching of opponents. The early dismissal of a star player, even if justified, denies spectators the chance of enjoying that player's skills and may lead to their team adopting boring defensive tactics for the remainder of the match.

The external disruption perspective

Whilst it is clearly the zone beyond the venue which provides most hazards perceived by the outside world, nevertheless the event area itself provides two main sources. First is the noise created by the participants or performers. This is a particular issue with music events staged in stadia, where the sound travels beyond the stadium through the open air.

Second is the threat of articles from the event area being projected beyond the facility. Cricket balls 'hit for six' out of the ground can damage property or injure passers-by. Pyrotechnics set off on the event area provide a further source of hazard. During an early satellite television broadcast from a Premiership football ground, some of the pre-match fireworks landed, still burning, on the forecourt of a petrol station down the street!

The safety/security perspective
The principal sources of hazards from this perspective involve perimeter obstructions and the potential for adverse interaction between spectators and participants in the event.

Television cables, perimeter hoardings, fences and gates all provide tripping or obstruction hazards which may delay spectator egress onto the event area in emergency evacuations. The 1989 Hillsborough stadium disaster in which 95 football fans were crushed to death against a perimeter fence provides an extreme example.

Incursions onto the event area are the other principal concern. The perceived hazards range from attacks on officials or players, for example the on-court stabbing of tennis star Monica Seles in Germany, to damage to the event area, such as at Wembley Stadium after a notorious England versus Scotland football match in the 1980s. Conversely, participants leaving the event area cause similar concerns. Players who run into the crowd to celebrate goals or points scored frequently cause the crowd to surge towards them. Players may even attack the crowd! Who could forget the pictures of Manchester United's Eric Cantona leaping into a stand to karate kick a spectator? Team benches provide a source of similar hazards, either because spectators misbehave towards them or vice versa.

Finally, we have the health and safety hazards which the event, the event area or the perimeter pose to the participants themselves. For safety reasons, the English football team nearly refused to play on a poor surface in China in summer 1996. More extremely, several boxers have died in the ring.

Thus from the event area alone, it can be seen that strategic risk assessment requires a broader focus than safety and security alone. Let me continue by looking at the next two zones, taken together. These are the viewing accommodation and inside concourse zones, including the various technological systems used to support their management.

Viewing accommodation and inside concourses

The commercial perspective
Design and fitting out are very much shaped by commercial risk concerns. Developers will want to recover as high a percentage of their capital costs as possible through advance sales of executive boxes and term tickets for premium seats. Operators will want to maximise the revenue streams from ordinary ticket sales by fitting as many seats into the facility as the various constraints will allow. Funding for many redevelopments has also been underpinned by the idea of diversifying the uses made of the facility, through conferences and banqueting, on days when there are no spectators in to view an event.

Factors which interfere with the opportunity for spectators to gain access to the event represent a further source of risk. Venues want to sell as many tickets as they can, yet police insistence on 'all-ticket' matches or their refusal to allow sales on the day have adversely affected attendances at some British football grounds.

Access to the event – yes, but free admission – no. Revenue protection means that entry to the inside concourses needs to be strictly controlled. This ensures that only those who have either paid or else been properly accredited are permitted to enter the facility.

Having got the audience in, merchandising seeks to address the risk that ancillary spend per head will not be optimised. More and better retail outlets, together with branded confectionery and catering items, increasingly seem to provide the answer here.

Risk may also arise from anything which increases costs or which prevents the maximisation of promotional opportunities. There are frequent tensions here between commerce and safety/security. Commercial managers will want the level of security personnel employed to be no more than is necessary to deal with the numbers and type of crowd expected, whilst regulatory agencies will be tempted to up the staffing levels 'just in case'. Commercial managers will want to earn revenue by allowing access to promotions, yet the promotional activity may itself compromise safety. At an old London football ground, a local publisher was allowed to place a free copy of his newspaper on every seat in the main stand, which happened to be made of wood.

Finally, there are concerns not to offend the occupants of executive boxes and damage repeat business by over-controlling their behaviour. Thus normal security personnel may be replaced by 'lounge stewards' who are encouraged to show more tolerance and tact than would be the case with the ordinary public.

The spectator perspective

The main areas of spectator-perceived risk concern the ease with which they can purchase the right to a seat or space and the quality of their enjoyment of the event.

Ticketing systems have become ever more sophisticated. Theatre box-offices are used to allowing the customer to choose exactly which seat they wish to purchase, but this is still a rarity in British sports. Yet there are still venues where even credit card sales over the telephone are not provided for and the prospective customer has to attend the venue in person to purchase the ticket.

Once they have gained access, spectators worry about whether they will be able to see the event. In addition to the viewing obstructions around the perimeter, the view quality is affected by three factors: preferred viewing location, viewing distance and sightline [9, 10]. The preferred viewing location for athletics is the side where the finishing line is. Rugby fans prefer the sides whereas younger football fans prefer the ends. For most team events, optimum viewing distance is a radius of 90 m whilst the accepted maximum is 150 m. Yet in several famous stadia, most spectators are beyond the optimum and far too many are beyond the maximum. Sightlines are assessed using riser heights, tread depths and angles of rake. Ideally, the spectator should be able to see over the head of the person in front, but this has often not been achieved.

Given a decent view, the spectator is then concerned with enjoyment. Risks here arise from failures in maintenance – such as dirty or broken seats, from poor amenities and, above all, from being prevented from having a good time.

Spectator amenities will mainly be sited in the inside concourse areas. Here the fans are looking for both ready access and a choice of quality in catering and souvenirs, as well as for sufficient clean and decent toilet provision. All too often, they face the risk of their loyalty to the sport or team being unscrupulously exploited. Long queues, foul latrines and over-priced insipid fare are still the norm in too many venues.

Enjoying the event is clearly key, and it is here that spectators may unwittingly come into conflict with the regulators. For some sports fans, being forced to sit in a designated seat rather than to choose where or even whether to sit, is an infringement of the right to enjoy themselves. The same is true of regulatory restrictions on the banners, flags, air-horns, drums and instruments which go towards creating the carnivalesque atmosphere which so contributes to the enjoyment of the live event. Risk for spectators also arises when security personnel respond to their passionate partisan support and letting off steam as though it was hooligan behaviour.

The safety/security perspective

A considerable proportion of the perceived risks arise and are addressed in the preparations for the event. Periodic inspections will be carried out on the structural integrity of the viewing accommodation, for example to check loadings, and extensive pre-event checks will take place to ensure that risk is reduced. There is a whole range of technological life safety systems – turnstile counting, crowd pressure sensors, lighting, closed circuit television, fire safety, communications and public address – which need to be working correctly to fully support the operational management of the event.

Furthermore, managers will want to be satisfied that the venue is clear of hazards and that all personnel are on post before they open the venue to the public. An England under-21 international football match at Wolverhampton was delayed for nearly three hours after a suspect object was found during a pre-match search of the viewing zone. Several stadium events have also been called off due to toilets or fire equipment having frozen.

In addition to its importance for revenue protection, access control is both a security and safety issue. Many venues are designed so that inside concourses and viewing accommodation together form self-contained areas, perhaps one for each side of the venue. Allowing too many people into an area creates a serious risk of overcrowding and possible disaster. Allowing unauthorised persons in compromises safety – if lots of people are involved – and security – if the person's intentions are sinister. At one London ground, a person walked through the players entrance dressed in a tracksuit and 'warmed up' with the teams until somebody realised he was an intruder.

Security risks arise from members of the crowd arguing over seat occupancy, committing criminal offences such as abusive chanting, throwing missiles, being drunk, fighting or reacting to the event with language or behaviour which is regarded as unacceptable by the authorities. Major crimes such as rioting, wounding or even unlawful killing are well known to have occurred in various sports. PAFs handle considerable sums of cash and several have been the victims of robberies.

Safety risks arise from areas of the viewing accommodation approaching capacity and from any factor which necessitates either a partial or total evacuation of the venue. These range from equipment failures – for example floodlighting – to fires, floods, gas leaks, explosions, bomb scares, structural collapses or serious public disorder.

Within the inside concourses, locked exit gates represent a particular safety hazard, since they prevent crowd egress in an emergency. Over 50 people burned to death in Bradford in 1985 and British football has learned this lesson. However, the same cannot be said of all other sports or countries. In 1993, I went to a cricket test match where I found the exit gates to a wooden stand locked. I was horrified to be told that the steward who held the keys was taking tea in the pavilion several hundred yards away. In October 1996, I visited a French football ground where all the exit gates were kept padlocked yet unmanned throughout the match.

The external disruption perspective

Since it is the viewing accommodation and inside concourses which form the bulk of the PAFs structure, it is the impact of the built form itself which provides the main source of risk to the world beyond the facility. This is less of an issue where PAFs are constructed on greenfield or redundant industrial sites. However, many PAFs are sited in cramped inner city locations and any redevelopment has to take account of the environmental impact on local residents. For example, the huge new stand at Dublin's Croke Park (the home of Gaelic football) was designed so that houses in the area would not lose sunlight either in the morning or the evening. Research carried out by Helen Rahilly has shown how, to meet the considerable objections of local residents, the final design of Arsenal Football Club's North Stand was "lower, lighter and far less bulky than the original plans had suggested" [11].

Noise and light pollution are further sources of risk. Light pollution occurs when the glare from floodlighting spills over onto surrounding properties, whilst noise pollution refers both to the noise of the crowd, which is perhaps unavoidable, and to the transmission of music and messages over public address systems which carry beyond the venue.

Let us now look at the principal hazards perceived in the final two zones – the outside concourses and the neighbourhood.

The outside concourses and neighbourhood

Since merchandising makes a valuable contribution to revenue, pirate merchandise represents a serious commercial threat for PAFs managers. Not only will no license fee have been paid but the merchandise itself may be very poor, damaging customer perceptions of quality and thus reducing official sales. Within the UK, brands are protected by copyright legislation and a number of venues have employed security personnel to patrol the environs to seize any pirate merchandise. This is sometimes undertaken in conjunction with Trading Standards officers who have the power to prosecute offenders.

Similarly, as PAFs look to earn more revenue from sponsorship and licensing arrangements, so ambush marketing becomes an important commercial hazard. James M Curl has described how ambushers can be 'locked-out' through the complete control of images in and around the venue [12]. The use of brand images in the venue environs can be controlled by local government permit whilst the outside concourses can be strictly patrolled to enforce restrictions on banners, signs and even clothing.

Spectators are rather less likely to perceive the wearing of a branded sweatshirt as a hazard. Outside the venue, their concerns centre around the ease with which they can gain access to the event and then how quickly they can get away afterwards. Perceived hazards arise from inadequate public transport, poor road capacities and difficulties in parking near the venue. Quite perversely, the same people who see their convenience threatened by parking restrictions before the event, will bemoan the absence of action against the parked cars which hinder the progress of traffic leaving after the event. Within the outside concourses, spectators wish to be guided effortlessly towards the right entrances and then gain admission without having to queue for more than a couple of minutes. Clear information on tickets, the best of signage and sufficient turnstiles or access points are essential if spectator frustrations are to be avoided. The needs of disabled patrons must also be met.

From a safety/security perspective, the principal hazards are perceived as disorder and overcrowding arising during the periods before and after the event. Particularly within football, there are public order concerns around spectators travelling to the match or gathering in public houses near the ground. Police have sophisticated intelligence systems which may cause them to believe there will be disorder and thus wish to make careful arrangements to provided supervised or even segregated routes to and from the stadium for the supporters of different teams.

The cramped inner city locations of many facilities create real risks of overcrowding in the surrounding streets, particularly as the start of the event approaches and many people are still outside waiting to gain admission. It is for this reason that the kick-off times of football matches are sometimes put back by 15 minutes and public address announcements made outside the ground to reassure fans they will not miss the match. Similar crowding risks arise where the venue is sold out and there are substantial numbers of people locked out of the event. Equally, at the end of the event, the exits onto the outside concourses will be opened in plenty of time and traffic stopped to ensure the crowds can disperse as freely as possible.

Within the outside concourses, supervised access control is perceived as essential to screen out drunken persons and minimise the risks of people taking dangerous articles such as flares, missiles or weapons inside the venue. For all-ticket capacity events, cordons may be placed at the boundary to the outside concourses, or even in the surrounding streets, to reduce crowding by restricting access to ticketholders only.

As far as local residents are concerned, the principal hazards arise from the nuisances created on the days when the venue is open for mass spectator events. These include the noise of the approaching crowd, the litter which the crowd leaves in its wake, minor disorder or vandalism, people urinating in the street or in front gardens, traffic congestion and the impossibility of finding a parking place near one's own house. Whilst these quality-of-life issues may be less keenly felt where venues are sited out of town or on derelict industrial land away from the main conurbation, nevertheless studies by geographers have shown how negative effects are experienced over quite some distance around a facility [13, 14]. Such widespread effects go some way towards explaining the "not in my back yard" campaigns so often mounted against PAFs developers seeking a site for a new facility.

IMPLICATIONS FOR PAFS MANAGEMENT

The constraints of space mean that I have highlighted only some of the principal areas of risk arising within each of the four perspectives. There are still considerable gaps in the analysis, which I acknowledge is skewed towards my own research in British football grounds. Many readers will undoubtedly be able to add considerably to what I have outlined.

But this is exactly my point. Having now looked in overview at the zones in the PAFs system, the full complexity of the balancing act required from facilities managers is becoming ever clearer. So what I want to do now is suggest a practical management approach which acknowledges the validity of different perceptions and thus allows for greater richness, diversity and consensus in the analysis. The approach involves the five stages of identifying, estimating, evaluating, managing and monitoring risks.

Hazard identification and risk estimation can both be carried out at a risk assessment workshop at which representatives of each of the four groups – commercial, regulatory, spectator and neighbourhood – can be invited to participate. The representatives should have a 'hands-on' knowledge of the issues and may therefore be relatively junior in status. Three representatives from each group would give a manageable workshop of twelve participants. Using the idea of the PAF as a system, with each zone broken down into a number of smaller areas, the group should be facilitated through the identification of any hazards which they perceive in each area. Participants should be assured that there are no right or wrong answers, that candour is welcomed and the validity of all views will be recognised.

This facilitated process will result in a comprehensive list of perceived hazards. Each hazard should then be considered by the whole group and a collective judgement made about its probability of occurrence. A second collective judgement should then be made about the potential adverse consequences if the hazard did occur. I would suggest that there are at least four types of adverse consequences to consider: for the profitability of the business (including its exposure to liability); for the enjoyment of the spectators or participants; for public safety and order; and for the community and environment in the outside world. These judgements about probability and consequences should be made using an appropriate scale upon which everyone in the group can agree. My preference is for a five point scale where 0 = None, 1 = Low, 2 = Low/Medium, 3 = Medium/High and 4 = High.

Thus the outcome of the workshop is a comprehensive hazards register which should provide a substantial reference document to support the operational management of the venue. But PAFs managers cannot reasonably be expected to tackle all the hazards, nor will it be cost-effective to try to do so. So the risks will need to be evaluated by an appropriate forum, which may well be the board of directors of the company running the venue, or their public sector equivalents.

One way of evaluating the risks is to multiply the probability and consequence ratings to give an overall risk evaluation rating. Using the five point scale outlined above would give a rating somewhere between 0 (0×0) and 16 (4×4). Whatever the chosen method, the general principle governing such evaluation is that risk should be reduced to a level which is 'as low as is reasonably practicable' [15]. Hazards which have been judged to be of lower probability and consequences will be designated as 'low risk' and will be accepted as tolerable. At the other end of the scale, hazards which have been estimated as higher probability and consequences will be designated as 'high risk' and therefore subject to remedial action, in some cases irrespective of cost, through the preparation of appropriate risk management plans. Where the boundaries fall between these two categories will be a question of management judgement, and, once decided, will determine which hazards are designated 'medium risk'. These may require careful monitoring with action where something can be done at a cost less than the benefit of the risk reduction.

The identification, estimation and evaluation processes should then be reported as a formal risk assessment. This documents which hazards have been identified as priorities and how and why those decisions have been made. So the risk assessment process helps PAFs managers in discharging their accountability for profitable, safe, enjoyable and minimally disruptive facilities management.

For each high risk or relevant medium risk, a second workshop should meet to consider what action can be taken to control the risk and reduce the probability and/or consequences to a tolerable level. Again, the workshop should comprise representatives of each of the four different groups, although it would now be appropriate for more senior people to be involved in the exercise. The countermeasures defined, the resources assigned and the responsibilities allocated should be recorded. The outcome of the process will be the risk management plan.

Having implemented the plan, regular monitoring is important to ensure that the implications of any changes are considered and appropriately acted upon. A formal review of the hazards, their estimated probabilities and consequences, tolerability and any risk management measures proposed, should therefore be carried out at appropriate intervals, for example after major building work. The hazards register, risk assessment and risk management plan should be amended as appropriate and reissued accordingly.

This approach can work not only for existing PAFs, but also for venues which are being renovated or even which have not yet been built. Engineering drawings can be used to determine the zones and areas in the proposed system and workshop representatives can be drawn from the groups who will eventually be involved in the facility. Carrying out a risk assessment should assist in early resolution of potential operating difficulties, thus allowing for the design to be changed to eliminate unnecessary operating costs, safety problems, external disruption and so on. For example, there may be a perceived hazard that personal radio communications will not work under a large stand because of the density of concrete. Identified at the design stage, this allows for leaky feeders to be built in during construction rather than as an expensive retro-fit.

CONCLUSION

We have seen how managing hazards and risks is an integral part of PAFs management. Four different groups give priority to different types of hazard in the five zones in the PAFs system. Commercial viability, excitement and enjoyment, safety and security and environmental impact are all legitimate perspectives on risk. The adoption of a more strategic and systems approach to risk assessment and management can help PAFs managers meet the challenge of striking an appropriate balance.

ACKNOWLEDGEMENT

This paper is derived from a series of articles first published in *Stadium and Arena Management* magazine. Earlier versions of some portions of the text were also included in the author's previously published work cited at references 6 and 7 below.

REFERENCES

1.   Wootton, G. and Stevens, T. *Into the next millennium: a human resource development strategy for the stadia and arena industry in the United Kingdom*, Swansea Institute of Higher Education, 1995.
2.   Elliott, D., Frosdick, S. and Smith, D. The failure of legislation by crisis. In Frosdick, S. and Walley, L. (eds.) *Sport and safety management*. Oxford, Butterworth-Heinemann, 1997. pp. 11–30.
3.   Wells, C. *Corporations and criminal responsibility*. Oxford University Press, 1993.
4.   Frosdick, S. Risk as blame. In Frosdick, S. and Walley, L. (eds.) *Sport and safety*

management. Oxford, Butterworth-Heinemann, 1997. pp. 33–40.

5. Frosdick, S. Cultural complexity in the British stadia industry. In Frosdick, S. and Walley, L. (eds.) *Sport and safety management.* Oxford, Butterworth-Heinemann, 1997. pp. 115–35.

6. Frosdick, S. Managing risk in public assembly facilities. In Frosdick, S. and Walley, L. (eds.) *Sport and safety management.* Oxford, Butterworth-Heinemann, 1997. pp. 273–82.

7. Frosdick, S. Sports and safety: leisure and liability. In Collins, M. and Cooper, I. (eds.) *Leisure management: issues and applications.* Oxon, C.A.B. International, 1997.

8. Mars, G. and Frosdick, S. Operationalising the theory of cultural complexity: a practical approach to risk perception and behaviours. *International Journal of Risk, Security and Crime Prevention,* Vol. 2, No. 2, April 1997, pp. 115–29.

9. Football Stadia Advisory Design Council. *Seating – sightlines, conversion of terracing, seat types.* London, The Sports Council, 1991.

10. John, G. and Campbell, K. *Handbook of sports and recreational building design,* Vol. 1: *Outdoor sports* (2nd edn.). Oxford, Butterworth Architecture, 1993.

11. Rahilly, H. *Environmental policy and practice in premier league football.* Unpublished BSc thesis. Liverpool Hope University College, 1996.

12. Curl, J. Ambush! *Panstadia International Quarterly Report,* Vol. 1, No. 2, Summer 1993. pp. 18 and 67.

13. Bale, J. In the shadow of the stadium: football grounds as urban nuisances. *Geography,* Vol. 75, No. 329, 1990. pp. 325–34.

14. Mason. C. and Robins. R. The spatial externality of football stadiums: the effects of football and non-football uses at Kenilworth Road, Luton. *Applied Geography,* Vol. 11, No. 4, 1991. pp. 251–66.

15. Health and Safety Executive. *Use of risk assessment within government departments.* Sudbury, HSE Books, 1996. p.16.

8 PREDICTION OF CROWD MOVEMENT USING COMPUTER SIMULATION

P. E. CLIFFORD
Halcrow Fox, London, UK

SUMMARY: This paper describes the application of pedestrian simulation modelling for crowd movement and pedestrian capacity analysis. The paper assesses the design process for sports facilities and other buildings and seeks to encourage a consistent approach as well as summarising the benefits that simulation modelling can offer in the design and safety assessment for stadia.
Keywords: Capacity, crowd, design, evacuation, modelling, movement, pedestrians, simulation.

INTRODUCTION

Prediction is often defined as to foretell or prophesy. Knowing how some prophets have been treated in the past it might seem courageous to suggest that prediction is the task in hand for pedestrian planning. Nevertheless, it has become evident from the increasing concentration of crowds at large sports events that current designs and the re-planning of existing facilities must consider crowd movement and facility capacity. The risks of ignoring such assessment procedures are high.

The process of predicting crowd movement requires a procedure that permits evaluation of designs without losing sight of the design elements. Traditional capacity analysis has used broad pedestrian and crowd demand assessment, combined with the manual calculation of capacity availability. J. J. Fruin [1] took the approach one stage further by defining levels of service (LOS) in relation to person density. The application of available space per person and overall passageway capacity has since been applied on a worldwide basis.

This paper explores the further development of the spatial assessment approach and considers the way in which computer simulation models have led to the wider application of pedestrian planning and can contribute to better design and operational assessment.

CROWD MOVEMENT

'Crowd' is defined as *a throng or mass (n.) or a group that flocks together (v.i)* as distinct from 'to crowd out', which has been defined as *to exclude by excess already in* [2]. Using

Stadia, Arenas and Grandstands, edited by P.D. Thompson, J.J.A. Tolloczko and J.N. Clarke.
Published in 1998 by E & FN Spon, 11 New Fetter Lane, London EC4P 4EE, UK.   ISBN: 0 419 24040 3

such definitions helps to capture the essence of the problem. A group of varying size that can move about and cause the physical exclusion of others from either the group itself or the space the group is in. Designers face the dual problem of trying to predict the capacity of a facility as well as the impact that the design may have on both normal and emergency operation.

THE ASSESSMENT PROCESS

In considering the assessment of design capacity the levels of service (LOS) approach [1] has defined categories which can be correlated with person comfort. The LOS that is comfortable for people will vary according to the environment or task in hand. Spectators are used to congested conditions and are thus far more likely to accept higher densities for the duration of sporting events. However, it is still necessary to reduce the risk of serious congestion in walkways, concourse and milling areas so that safe passage can be maintained. As density increases it is likely that people will become more agitated and prone to reactive behaviour. The delay to walking progress on any route can increase the level of frustration for individuals and it is thus particularly important for the design process to consider the likely delays as well as the theoretical capacity. Computer simulation has enabled the pedestrian planner to re-calculate levels of service and person delays for very short time periods and thus provide a dynamic assessment of crowd movement through facilities.

Models include different passenger assignment techniques, which can be used to simulate different behavioural and physical layout scenarios. These include fixed routes, defined by the user, and dynamic assignment techniques, such as equilibrium or stochastic assignment algorithms. The equilibrium process seeks to minimise the route times for pedestrians. The stochastic assignment technique assumes that a proportion of people seek other than the optimum route. Validation work on a number of projects has found that the stochastic assignment technique provides a good simulation of route choice behaviour.

The route choices made by models [3, 4] consider the origins and destinations (locations) available to the pedestrians. These locations are defined by the user, as are the links between the locations. During the process of simulating the passenger demands through the building layout simulation models can consider the walking speeds and delays to pedestrians and summarise the pedestrian flow and occupancy for each area during user defined time periods. The flow from one area to another depends not only on the layout, but also on the number of people in that area and the distances involved. The PEDROUTE and PAXPORT models assume that the time taken is a function of the density of passengers in each area with two types of relationship being defined:

- function of time to density (e.g. for concourse/milling areas)
- speed/flow function (e.g. for passageways/stairs).

The rates of flow from one area to another are then limited by the flow rate for a given area and by the density of people in that area. The maximum density acceptable in any area of the facility can be defined by the user. If it is exceeded the model will scale down the flows able to enter that area.

Further to the process of individual person modelling, procedures for the allocation of group sizes, where persons move together, have been defined in simulation models. The models can also consider the space being used in person equivalent units, originally developed to assist in the evaluation of the use of trolleys at airports.

The process of simulation modelling has been applied to facility design for railway stations [5], airport terminals and buildings [6] and has used the LOS and delay approach to predict the likely densities of crowds as well as to estimate the time it will take to clear the facility in both normal and emergency scenarios.

APPLICATION

To apply the person density category analysis developed by Fruin [1] it is necessary to investigate the acceptable LOS for each area of a facility. This is unlikely to be the same for each facility or group of people. However, good design needs to address the issue of density in spatial areas affecting the time for crowds to progress to/from their stands or seats in stadia, as well as the basic LOS. Here the simulation model, which seeks to forecast the progress of individual persons through the system, is invaluable.

Software has been developed for the assessment process, including the PEDROUTE [3] and PAXPORT [4] programs, which were originally developed with London Underground Limited and BAA plc, respectively. These models grew from the desire for design teams to understand the passenger capacity performance of transport interchanges. Since their early development in the 1980s the models have been enhanced to consider route choices and person behaviour and are being continuously updated as architects, engineers and planners seek more detailed model output. The current PAXPORT model considers person types and can model the different routes chosen, the different times to perform tasks (such as at payment booths) and the dwell times at other facilities. Both the models listed above can also simulate train service provision at adjacent railway interchanges, which are often a major part of the design assessment process for sports facilities.

Fruin [1] developed level of service categories A to F, with A representing free flow conditions and F highly congested conditions. The LOS category analysis described can be broadly summarised as follows:

Category	Description
A	Free Flow conditions
B	Minor conflicts only
C	Some Restrictions to speed
D	Restricted movement for most
E	Restricted movement for all
F	Shuffling movements only

The person densities relating to the LOS vary for the area defined, with Table 1 illustrating the metric conversion of Fruin's [1] categories.

Whilst the LOS category analysis described is not new, the regular and consistent application of the procedure is becoming more common as computer simulation permits the calculation of LOS levels on a dynamic basis rather than as a one-off calculation. The combination of the adjustment of capacity through passageways and stairs, using speed/flow

Table 1. Person densities (square metres per person) related to the Level of Service category.

Level of Service (LOS) category	A	B	C	D	E	F
Walkways	>3.25	3.25–2.32	2.32–1.39	1.39–0.93	0.93–0.46	<0.46
Stairways	>1.86	1.86–1.39	1.39–0.93	0.93–0.65	0.65–0.37	<0.37
Queueing areas	>1.21	1.21–0.93	0.93–0.65	0.65–0.28	0.28–0.19	<0.19

curves and time/density functions has enabled models such as PEDROUTE [3] and PAXPORT [4] to adjust walk times in relation to opposing flows, width of the facility and route choices being made. The additional ability of simulation models to consider the impact of congestion downstream has also enabled models to consider how congestion is likely to build up within a facility as pedestrian demand varies. The current models use very detailed time slices (second by second) although model results are often reported in 1–5 minute periods for ease of presentation.

SCENARIO TESTING

With the concept of good design comes the issue of fitness for purpose and the safety for users. Sports facilities are required to serve the programmed events and to cope with heavy ingress and egress demands. Whilst these high demand levels pose design capacity issues, the emergency evacuation scenario can be the greatest problem. During an evacuation time is of the essence.

Methods for calculating the ability of a facility to cope with an evacuation have developed from the manual application of width of walkway, capacity available and walk times along a route. The original manual procedures do not consider the dynamic build up and decay of queues, which can cause serious delay and turbulence to free flow movement. The computer simulation approach redresses the technical balance and can provide output on:

- levels of queues, and
- delays at pinch points in the system.

The above results can be output for either normal operation or emergency evacuation scenario tests, along with the total time taken to clear the facility.

The benefits of a dynamic computer simulation of people at sports events include the ability to:

- test different event types (sports, concert etc.)
- test different visitor numbers
- test different layouts
- test evacuation times
- assess public transport interchange
- assess the impact of retail facilities on-site,

FUTURE DEVELOPMENTS

Current models have introduced the detailed simulation of complex pedestrian flows on facility layouts. The presentation of the results has progressed from data files containing large quantities of figures to the graphical representation of category analysis (LOS), delay and flow data by hot-spot colour systems. Additionally, smaller area and time slice information is being displayed using visualisation techniques and progress is being made toward more dynamic virtual reality displays. The latter permit the display of results for qualitative assessment, although the quantitative assessment using person densities by time period, capacity/demand ratios and route walk time still provide the most consistent form of design assessment. The dynamic display of individuals gives the design team a feel for the numbers of people present in each area, but the final decisions in the design process require a quantitative process that addresses the main issues of capacity over time against dynamic demand.

ACKNOWLEDGEMENTS

The author would like to thank Niels Hoffmann of Halcrow Fox, who has been responsible for the development of the PEDROUTE and PAXPORT models, for his assistance in writing this paper. The views expressed in this paper are not necessarily those of Halcrow Fox.

REFERENCES

1. Fruin, J. J. *Pedestrian Planning and Design.* Metropolitan Association of Urban Designers and Environmental Planners, 1971.
2. *Collins English Gem Dictionary.* Collins, 1968.
3. Halcrow Fox. *PEDROUTE 4 for Windows User Guide*, Vols 1 and 2. Halcrow Fox, 1997.
4. Halcrow Fox. *PAXPORT User Guide.* Halcrow Fox, 1996.
5. Clifford, P. Passenger simulation modelling in station planning and design. Computers in Railways V. Vol. 1: Railway Systems and Management, *Proceedings of Comprail 96 Conference, 1996.* Vol. 1 pp. 229–36.
6. Bulman, E. and Clifford, P. Customer service - interchangeable? *Passenger Terminal World*, July-September 1996, pp 18–23.

9 MINIMISING THE EFFECTS OF TERRORISM BY DESIGN

C. J. R. VEALE
Government Security Advisor, UK

SUMMARY: Certain structural and architectural measures can minimise injury to people and damage to assets in the event of a terrorist outrage. By including these measures at the design stage, not only are they more effective but they are also better value than retro-fit solutions.

The main areas of concern are: the provision of robustness; the incorporation of redundancies; the protection of 'key elements'; facade design; internal planning; and glazing protection.

Keywords: Terrorism, explosion, blast wave, primary and secondary fragments, robustness, key elements, stand-off glazing protection.

INTRODUCTION

Terrorism impacts on our daily lives and all citizens have a responsibility, in whatever way we can, to minimise that impact. However, with the influence of European law – and especially with the legal requirements of the Health and Safety at Work Regulations 1992 (Section 7: *Procedures for serious and imminent danger*) – an owner/occupier has an explicit responsibility for the safety of staff and visitors. In the event of a potential terrorist threat to or at an establishment, it is the owner/occupier or his representative who has the responsibility for making decisions regarding safety, not the police. The police will offer advice but they will not take responsibility. By incorporating certain measures into the design the owner/occupier is not only able more easily to fulfill his responsibilities regarding safety but also to reduce the potential for injury and death.

BACKGROUND – THE THREAT

Recent Provisional Irish Republican Army (PIRA) bombing campaigns on the United Kingdom mainland have been primarily against prestigious commercial and industrial targets and the transport infrastructure. The establishment – individuals and locations connected with government and the Armed Services – was becoming better protected, and preventative measures were being steadily applied, reducing the chances that attacks would be successful. Prestigious commercial and transport targets offered PIRA the opportunity

Stadia, Arenas and Grandstands, edited by P.D. Thompson, J.J.A. Tolloczko and J.N. Clarke.
Published in 1998 by E & FN Spon, 11 New Fetter Lane, London EC4P 4EE, UK.   ISBN: 0 419 24040 3

to gain publicity and to inflict severe economic damage to businesses and the country, in part through the consequences of the attack itself and in part through the possible reaction of future investors, the travelling public and indeed through the impact on public opinion at large.

Terrorism is about the infliction of terror and the general public is usually its target, either directly, when victims of attacks are in public places, or indirectly, when public order and governance appear to be undermined. PIRA is not the only terrorist group, and the tactics it has adopted will prove attractive to others. Islamic extremist terrorists, for example, pose a threat to western interests. Attacks against the Israeli Embassy and Balfour House in London in 1994 were directed explicitly against Jewish targets, but Algerian terrorist attacks in France in 1995 and 1996, especially those on the Paris metro, illustrate how the public and the economy may be seen as prime targets for a campaign of terror. So why should we consider designing out terrorism?

Structures generally have a long design life, anything from 25 to 100 years. Terrorists' campaigns will come and go over such a period. History shows that peace will be made, and broken. Terrorist groupings will fragment, disappear and re-emerge. New causes and new crusades will emerge. The one constant of the past 20 years or more has been that the bombing attack is a feature of terror campaigns worldwide. There is no reason to suppose that this phenomenon will change. Structures designed and built now must thus contain as many features as possible to minimise the chances of death and injury and loss of significant assets not just next year but in 20 or 50 years' time.

THE MAIN PRODUCTS OF A DETONATED EXPLOSIVE DEVICE THAT CAUSE DAMAGE TO PEOPLE AND ASSETS

In simple terms, an explosion is the sudden release of energy caused by a very rapid chemical reaction which turns a solid into heat and gas. This reaction takes place in less than a milli-second and is described as the rate of detonation. In the process of turning a solid into a gas, expansion occurs. In an explosion the expanding gas is produced extremely rapidly and pushes the surrounding air out in front of it, creating a pressure wave, known as the blast wave. For modern military plastic explosives (PE), of which Semtex is but one type, the rate of detonation is of the order of 7 to 9×10^3 m/s. For the more common varieties of terrorists' home made explosives (HME) the rate of detonation is lower: 4 to 6×10^3 m/s. Although not as efficient as PE, HME is still a very effective explosive, as can be seen from the damage caused by the attacks in London and Manchester.

Terrorists will generally use HME in the large devices, e.g. cars, vans and lorries, because the constituents are cheap and freely available. PE, such as Semtex, has to be acquired from states which sponsor terrorism or the black market and can be relatively expensive. Therefore PE is generally used only in small quantities in specifically designed devices.

When an explosion occurs at ground level there are several effects created that cause damage and injury. The extent of these effects will be dependant on the power, quality and the quantity of explosive material deployed.

The six basic effects are:

- The fire ball
- Brisance or shattering
- Primary fragments
- Ground shock
- The blast wave
- Secondary fragments.

The fire ball
The ball of fire created as part of the explosive process: it is very local to the seat of the explosion and is very short lived (a few milli-seconds).

Brisance
Brisance is the shattering effect, it is very local to the seat of the explosion and is generally associated with high explosives.

Primary fragments
These are fragments of the device or its container which have been shattered by the brisance effect and are propelled at high velocity over great distances.

Ground shock
Ground shock is produced by the brisance effect of the explosion shattering the ground local to the seat of the explosion, i.e. creating the crater. The shock wave resulting from the crater's creation continues through the ground, and is known as ground shock.

The blast wave
The blast wave is a very fast-moving, high pressure wave created by the rapidly expanding gases of the explosion. The pressure gradually diminishes with distance.

Secondary fragments
These are fragments that have been created by the blast wave imparting pressure onto friable materials which are unable to withstand this sudden over pressure. Typical friable materials which form secondary fragments are glass, roof slates, timber and metal frames. The energy from the blast wave imparted to fragments of friable material can throw them large distances and at great speed.

METHODS OF ATTACK AND DAMAGING EFFECTS

The five most common current methods of attack against prestigious, commercial and industrial targets used by PIRA and other terrorist groups are:

- The telephoned bomb threat (TBT)
- The delivered item – postal, courier, etc.
- The improvised incendiary device (IID)
- The improvised explosive device (IED) – small man-portable devices

- The vehicle-borne and the large vehicle-borne improvised explosive device (VBIED and LVBIED) – devices concealed in cars, vans or lorries.

Of the above five, two – the IED and the VB&LVBTED – can have significant effect on the structure and architecture and it is these that we need to consider when attempting to minimise the effects of a terrorist attack.

IEDs

For a small man-portable device to cause any significant damage to a structural component, it will need to be touching or in very close proximity. Glazing, on the other hand, may be damaged at distances of up to 30 m, if the device is external. However, if the device is deployed internally then the containment effect and the blast wave reflections can increase that distance to 60 m.

V&LVBIEDs

The brisance effect and the blast wave from a large vehicle-borne device can cause severe damage to structural components up to 75 m from the seat of the explosion. Damage by the blast wave to architectural components, e.g. glass (secondary fragments) may occur up to 250 m away. Small primary fragments, the size of a 50 p piece, can travel up to 500 m, whereas the larger primary fragments, e.g. the engine block or rear axle may travel up to 80 m. However, if a large vehicle-borne device can be deployed inside a building or structure then the damage effects are very much greater as the blast energy is contained, i.e. there is no or limited venting of pressure, and there can be large increases of blast wave reflections.

Therefore it can be seen that by increasing the distance, often described as the stand-off, between the device and the target or vulnerable item, the damaging effects are significantly reduced. The damage reduction is a function of the distance cubed.

MINIMISING THE EFFECTS OF AN EXPLOSIVE ATTACK

There are four key areas that require attention:

Key elements (e.g. tension cables, anchor supports for cantilever structures, large span roof trusses)

The standard principles for robustness apply. That is, the key element has to be designed in such a way that either it is able to resist and not be compromised by all the products of an explosive device or if it is damaged by an attack there is sufficient integrity and strength remaining in the element to enable it to perform its designed function. This is particularly critical for tension members.

However, to achieve a design solution that can withstand the very onerous load cases arising from the blast wave loading and the fragment impact damage would require a massive and uneconomic structure. Therefore a better and more achievable approach would be to reduce the blast and fragment load cases by maximising the stand-off. This may be achieved by:

- Preventing access of unauthorised vehicles. This can be achieved by installing and operating good access control procedures; planting trees, erecting bollards, street furniture, etc.; 'hard' landscaping, incorporating ramps, steps, mounds, etc. and 'soft' landscaping. Care should be taken, however, when using 'soft' landscaping to ensure that concealment places for smaller devices are not created in the dense foliage.
- Preventing public access to the key element. There are several ways in which this can be achieved; by locating the element and its fixings out of reach; concealing it with an architectural feature; protecting it so as to increase stand-off.
- Designing the key element and its fixings such that it is impossible to deploy a device. Examples of this are: incorporating curved or sloping surfaces; using non-magnetic materials or surface materials; shrouding with a sloping surface covering.
- Designing the key element, its fixings and surroundings such that by a simple visual examination it can be quickly determined that nothing suspect has been deployed.

Façade detailing

There are several aspects of the façade design that should be considered when attempting to minimise the vulnerability of people within structures and the damage to the facade itself:

- Minimise the amount of glazing. This limits the amount of internal damage from the glazing and the amount of blast that can enter and cause damage to the internal fixtures and fittings.
- Ensure that the cladding is securely fixed to the structure with easily accessible fixings. This will allow rapid inspection and if necessary replacement after an event.
- Ensure that the cladding system allows for the easy removal and installation of any one panel. This will prevent the need to remove all the panels if only one is damaged after an event.
- Avoid the use of deep reveals and deep flat window sills which are accessible to the public, as these provide ideal concealment places close to the structure for the small devices.
- Minimise the use of deep surface modelling as these can enhance the blast effects by reflecting, the blast wave and causing disproportionately more damage than would be the case with a plane facade.

Internal planning

When designing the internal layouts ensure that there is a secure boundary between the private and public parts, including the installation and operation of good access control procedures at access points between the two parts. This will minimise the deployment of IIDs and IEDs in the private areas and allow the searching regimes to concentrate on the public areas

To assist the speedy searching of the public areas, minimise, at the design stage, the number of potential concealment places and ensure 'good housekeeping' procedures are followed, i.e. keep the public areas as clear and as clean as possible; do not use the public areas for storage; keep dust bins etc. in the private areas.

Glazing

In terrorist explosive incidents more than 90% of injuries sustained are from falling or flying glass. Both annealed and toughened glass, when subjected to the blast wave, smash into sharp shards, which are either flung into the structure or pulled out onto the surrounding area.

Blast resistant glazing, which contains the shards, generally consists of laminated annealed or toughened glass. The preferred minimum thickness for the laminated glass is 7.5 mm, which includes a minimum thickness of 1.5 mm plastic interlayer. This interlayer must be polyvinyl butyral for the glazing to offer blast resistance. To achieve the full benefits of the blast resistant glazing, the glass should be installed in robust frames which are securely fixed to the surrounding structure.

10 THE ROI BAUDOUIN AND STANDARD STADIA, BELGIUM: A CONCEPTION OF SAFETY AND ECONOMY

L. DEMORTIER and A. DUMORTIER
Engineering Office Greisch, Liege, Belgium

SUMMARY: The adoption of more and more stringent security standards, along with the joint organization by Belgium and the Netherlands, of the Euro 2000 competition, lead to the renovation of several large Belgian stadia. This paper describes the works to modernize the Roi Baudouin stadium in Brussels (50,000 seats) and the Standard stadium in Liege (30,000 seats) and discusses particularly the conception of these stadia, mostly lead by security and economic requirements.
Keywords: conception, cost, economy, safety requirements, multi-purpose stadia, renovation, security.

INTRODUCTION

The Heysel tragedy and its consequences

The final of the European Champions League between Liverpool and Juventus (Torino) was played on May 29 1985 in the Heysel Stadium in Brussels. The tension between some supporters led to fights and tragedy followed with 39 dead and 370 injured people.

Numerous factors allowed this catastrophe to happen (bad preparation of the match, disorganization of the security services, and so on) but the structure itself has above all to be incriminated: fences gave way, allowing the supporters of both clubs to be present in a same block; a wall broke, causing supporters to fall several metres.

After this tragedy, several commissions have been created to prepare guidance on football stadia [1], far more severe than previously. Numerous stadia, mostly those used for international and first division matches, including of course the national stadium, have then had to be either adapted or rebuilt.

The organization of Euro 2000

Belgium and the Netherlands proposed jointly to UEFA several stadia with well defined capacities:

- a minimum of 50,000 seats for the opening match and the final
- a minimum of 30,000 seats for the other matches
- always a minimum of two-thirds covered seats.

Stadia, Arenas and Grandstands, edited by P.D. Thompson, J.J.A. Tolloczko and J.N. Clarke.
Published in 1998 by E & FN Spon, 11 New Fetter Lane, London EC4P 4EE, UK.   ISBN: 0 419 24040 3

Several sound Belgian stadia do not yet match these requirements. Before the year 2000, they will then have to be enlarged, adapted or entirely rebuilt.

Multi-purpose stadia: football, concerts, etc.

Due to the numerous television transmissions, the multiplication of the matches and the limited interest of the Belgian public, the number of spectators per match has decreased, along with the financial income of the clubs. On the other hand, the subventions of the public authorities are limited.

Stadia managers increase the income from their infrastructures with a diversification of their use, such as the organization of concerts.

THE ROI BAUDOUIN STADIUM IN BRUSSELS

After the first stage of renovation works, the Heysel stadium has been renamed Roi Baudouin, after the late King. This stadium is to be used for athletics competitions, such as the Van Damme Memorial, as well as for field sporting events, such as football.

First stage of works (August 1994–August 1995)

These works were carried out in order to meet the new security standards. Three stands (out of four) had to be demolished and entirely rebuilt, creating 28,000 new covered seats, in a total of 40,000 seats in the stadium.

The frontage of the main entrance had to be preserved and protected. The structure of the facade overhangs it with a 40 m-long shell beam.

Second stage of works (January 1997–August 1998)

After the end of the first stage, Belgium was designated to organize the Euro 2000 competition. The capacity of the stadium then had to be increased to 50,000 covered seats.

Various solutions were studied, all based on the reconstruction of the fourth stand (Fig. 1):

- a huge stand, with three levels of stepped rows of seats (1)
- a huge stand, with a long second level in cantilever (2)
- a stand with a similar cross-section to the circular ones and the construction of a complementary level on these three stands (3).

The last solution, the least expensive, has been chosen because the overall harmony of the three stands improves the aesthetics of the stadium. The metal roof of the circular stands has therefore to be taken off and then put back.

Use of the stadium could not be interrupted during the works: all the international matches are played and the first concert in the stadium has happened. The works have then to be done in successive stages and each stand is completed in 6 to 8 months.

The completed stadium

The Roi Baudouin stadium, entirely rebuilt, will contain 50,000 covered seats, changing-rooms, large multi-purpose halls, two smaller sport halls, an 80 m-long covered running-track and offices for the national associations of football and athletics, and for the Van Damme Memorial organization.

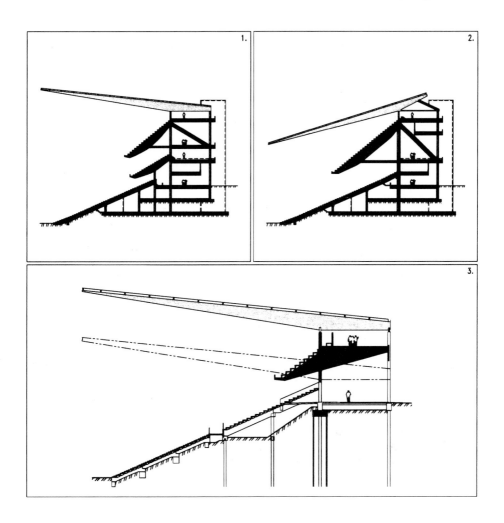

Fig. 1. Roi Baudouin Stadium in Brussels.
(1) (2) Not retained concepts of stand. (3) Extra level built above first stage stand.

The natural soil has the shape of a 12 m-deep basin, allowing the external height of the stands and, therefore, the costs to be reduced.

The structure of the stadium has to be as economical as possible and so has been mostly realized with exposed concrete. The load-bearing columns have a 1200 × 450 mm section, on a 11.4 × 7.0 m grid. The beams under the upper tiers have a 7.8 m-long cantilever.

The upper tiers (slope 51/75) have an equivalent width of 190 mm and the lower ones (slope 30/70) of 160 mm for a 7.2 m-long span. Their slope matches all visibility requirements. Each upper compartment is served by a minimum of three stairways.

The indoor running track is located on the upper passage-way and, for the comfort of the athletes, has a first mode eigen frequency greater than 11 Hz.

The roof, light and simple, is made of metal box-girders of variable height (200 to

2600 mm), 700 mm wide, located every two spans (14.4 m). All seats are covered with a 40 m cantilever, up to 30 m above the field. Three stands are covered with metal cladding (23,500 m²), and the main stand is covered with translucent polycarbonate sheets (4,500 m²).

The finite element method and a perfect knowledge of the metal structure design rules have allowed the design of the roof beams with a slenderness ratio of 300 and a stiffener every 6 m.

THE STANDARD STADIUM IN LIEGE

The Standard is the stadium of a first football division club. It has 20,000 seats, partially covered. The capacity has to be increased to 30,000 seats for Euro 2000.

This enlargement is realized by an extra level over two of the existing stands (Fig. 2) and over a new third stand (3,000 seats), which no longer meets the security requirements.

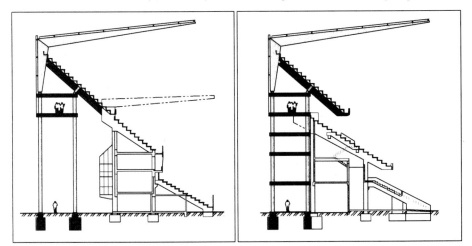

Fig. 2. Standard Stadium in Liege. Two existing stands with an extra level.

Composite columns, made of circular metal tubes (530 mm diameter, 5.5 mm thick) filled with concrete, are placed on a 6.00 × 5.85 m grid and carry a 338 m-long passageway, 16.2 m above the ground. Direct access is permitted by eight stairways, acting also as stabilizing elements under horizontal loads, such as wind or crowd (5% of the vertical loads). The stairways are made of double flights, in order to limit the overall dimensions and the cost.

Above the passageway, a triangular structure, made of floor beams, beams bearing the tiers and metal box-girders acting as prolongation of the roof beams, bears the tiers and the roof. The beams supporting the tiers, partially over the existing tiers, are tied on one end by a strut, in order to increase the frequency of the first vibration mode. The new roof (8,000 m²) covers both old and new stands with a 27 m-long cantilever, up to 31 m above the ground level.

The important slope of the tiers (31°) is due to visibility requirements. Security is improved by the presence of handrails in all the circulating ways. The tiers, 700 mm deep and 600 mm high, have an equivalent height of 140 mm for a 6 m-long span and are made of reinforced concrete. All the visible faces are formed and do not need any extra work.

The new tiers have 10,000 extra seats, offering four compartments with less than the maximum 3,000 required seats. Each compartment has two or three outlets, depending on the distances to walk to the exits.

The zone between the ground and the passageway stays free on two stands and is used as a circulation area around the stadium. In the third stand, four levels of offices and stores are created. These complementary levels, bearing on the main columns, are run totally independently, except for the emergency outlets that join the new stairways of the stand.

The works, expected to last a maximum of 15 months, began in January 1998 and are to be carried out without interruption to the scheduled matches. Finally, the stadium will present an overall harmonious and unified appearance.

SECURITY

Context
Belgium, due to its central European position, frequently attracts numerous foreign supporters during international matches, and, consequently, numerous 'hooligans' as well.

As Belgian laws are less severe than British ones concerning the penal prosecutions for violent actions in stadia, by their design conception, the stadia have to reduce the opportunities for violence and to resist such acts.

Constraints
Security requirements have became more and more exacting and, for Belgian stadia, come from various sources, the Sportive Events Security Commission, UEFA, the fire authorities, etc. Here are the important requirements concerning the conception of the structure.

- *Outer fence.* The stadium has to be surrounded by a fence that cannot be crossed.
- *Inner fence.* As barriers are used, exits have to be provided, with doors opening towards the field, even under the pressure of the crowd.
- *Stairways and outlets.* Their minimum width is the number of people that have to go through that passage, expressed in centimetres. This width has to be multiplied by 1.25 if the evacuation goes downwards and by 2 if it goes upwards. This width can be reduced with regard to the non-combustibility of the structure. For instance, for a stand totally made of non-combustible materials, the width can be reduced by a factor of 5. The minimum width is 1200 mm.
- *Pressure on the barrier railings.* A minimum horizontal pressure is required on the barrier railings, up to 5 kN/m, depending on the width of the passages in front of the railings.
- *Compartments.* The number of supporters is limited to 3,000 seats per compartment, with a minimum of two outlets.
- *Emergency outlets.* No spectator can be more than 30 m away from the nearest exit towards the evacuation ways.
- *Seats.* Their minimum width is 500 mm. The maximum number of seats on a single row

is 20 or 40, depending on their accessibility from one or both sides.

- *Stairways.* Maximum 17 steps par flight, maximum slope of 75%, straight stairways, minimum 250 mm long and maximum 200 mm high steps.
- *Railings.* They have to be provided in every stairway. Over 2.4 m wide, stairways have to be divided by railings.
- *Fire resistance.* Minimum 1 hour.

Circulation

- Access to the seats from outside the stadium is easy, direct and fast. Labyrinths are avoided.
- If possible, entrance doors are narrow (usually 800 mm) in order to allow easy control of tickets and searching operations by the security forces.
- At ground level, the exit doors, wide and with no obstacles, are located directly in line with the evacuation ways coming from the tiers.
- On the other floors, a passageway leads the spectators to the nearest stairway.
- The stadia can be entirely evacuated in less than 5 to 6 minutes.

Barriers and railings
All the outer barriers are able to resist to a crowd pressure of 5 kN/m, applied one metre above the floor.

Compartments
The use of barriers to separate compartments has several disadvantages:

- The supporters can clutch at each other through the bars.
- The visibility is really poor for the spectators seated along the barriers (up to half the field may be invisible).
- The reduced visibility makes the spectators more agitated, leading to the possibility of violence.
- The bars could help people to cross the barriers.

Glazed compartments allow the reduction or suppression of those drawbacks. They have a lower concrete wall, usually one metre high (in order to resist the horizontal crowd pressure and not to obstruct the view) with an upper glazed wall which is of high resisting class.

Security is increased if possible by providing stairways on both sides of the wall, to allow the circulation of the security forces.

Inner fence
The inner fences, separating the field and the contingent track from the tiers, are usually made of metal grating. The spectators feel then like being in a cage and their behaviour can reflect this.

In the Roi Baudouin stadium, an experimental trench (Fig. 3) is being included in order to avoid the use of fences. Only glazed emergency doors towards the field will remain.

Fig.3. Experimental trench.

The eigen frequency of the structure

Stadia no longer have a unique purpose. The structures have to withstand various vibratory effects with no damage or discomfort.

The main reasons for vibrations are synchronized responses of spectators (clapping of hands, stamping of feet, etc.) to music (hard-rock concerts and so on) and organized crowd moves (waves, jumps, etc.). The eigen frequency of the various structural elements has to be different enough from the eigen frequencies of the crowd actions. Analysis of the literature [2, 3] has led us to require as a minimum a 6.5 Hz vertical eigen frequency and a 2.5 Hz horizontal frequency.

DESIGN/CONCEPTION

Structures

The load-bearing structures, apart from the roofs, are made of concrete. This material has been chosen as well for economical reasons as for its good characteristics towards fire and combustion.

The roofs are made of metal box-girders, more elegant than trusses, cleaner (no access for pigeons) and easier to maintain.

Cost

The small amount of money available for modernizing the stadia required priority to be given to the load-bearing structure and not to architectural features. For the same reason, the formed concrete stays naked, in order to reduce the completion works (painting, etc.).

Here are the costs of the two stadia are as follows:

- *Roi Baudouin Stadium.* Overall cost: 1,395 million BEF for 50,000 covered seats, i.e. 28,000 BEF (£465) per seat.
- *Standard Stadium.* Total cost: 300 million BEF for 13,000 extra seats, i.e. 23,000 BEF (£385) per seat.

Delays

Construction periods are always kept as short as possible in order to get the stadia back to full operational capability as quickly as possible. The stadia are used during the works, with a limited capacity, of course.

These short delays require as much use as possible to be made of prefabrication, allowing concrete elements to be constructed even during the winter, and the aesthetics and quality of finishes to be taken care of continuously. Prefabrication is favoured, even for small elements. The tiers have been realized together with the steps of the stairways.

All beams, floors (hollow tiles), tiers, stairways and stair-heads of both stadia are precast. For the Roi Baudouin stadium, even the columns and the shells are precast; therefore, only the foundations and blindings have been concreted on site.

Standards
The most recent standards have systematically been used (Eurocode 2, Eurocode 3, railings standards, etc.).

CONCLUSION

The new Belgian stadia prove that small budgets can be harmonized with aesthetics and efficient functioning, and are fully compatible with the increasingly strict safety standards.

REFERENCES

1. Commission for security during the sporting events. *Manuel pour la sécurité dans les stades de football.* Home Office. Belgium. Edition 1991.
2. CEB. *Vibration problems in structures, practical guidelines.* Information report nr 209. August 1991.
3. Reid, B. *Design of Stadium Structures - A Safety First Approach.* AIPC. 1994.

PART THREE

STRUCTURAL DESIGN AND CONSTRUCTION

11   PRECAST CONCRETE IN STADIA

A. C. CRANE
Bison Precast Concrete Products Ltd, Tamworth, UK

SUMMARY: For speed of construction in the limited space of a stadium site, the use of precast concrete units is invaluable. This paper highlights the advantages of precast elements and describes the manufacturing process with particular emphasis on flooring, terracing and staircase units. The use of off-site fabrication and 'just in time' delivery methods ensure efficient construction programmes.

Keywords: Bearings, benefits of use, design/detailing, floors, football, precast concrete, stadia, staircases, terrace units, vomitories.

INTRODUCTION

The reasons for the impetus in stadia construction, particularly in the UK, are well known. The disaster at Hillsborough in 1989 and the Taylor Report that followed triggered a massive review and financial investment in places of large public assembly such as stadia. The largest expenditure has been on football stadia but there have also been major developments for other sporting venues such as rugby, cricket and horseracing.

Fundamental design guidance has been issued by a number of bodies such as the Football Association and the Health & Safety Executive. First and foremost, these guidelines have been set out to ensure the safety of the public but there has also been a real intent to raise the quality level of the accommodation provided and recommendations have been made on matters such as lines of sight, seat spacing (both laterally and with regard to leg-room), means of access and egress, toilet facilities etc.

A great deal of progress has been made in designing stadia for wider, multi-purpose use. The days of a stadium occupying a valuable city site and generating a significant revenue on alternate Saturdays for eight months of a year cannot be justified in today's society.

The Cardiff Millennium Stadium is a prime example of a stadium to provide for today's and, hopefully, tomorrow's requirements.

BENEFITS OF USING PRECAST CONCRETE

The benefits of using precast concrete in many types of stadia include the following:

Stadia, Arenas and Grandstands, edited by P.D. Thompson, J.J.A. Tolloczko and J.N. Clarke.
Published in 1998 by E & FN Spon, 11 New Fetter Lane, London EC4P 4EE, UK.   ISBN: 0 419 24040 3

- Off-site production
- Control of quality
- Speed of erection
- Immediate working platform
- Minimum of propping
- Minimum of wet trades.

Generally, these can be regarded as common to most types of structure, but it could be argued that, in many ways, stadia work is an application where some of the benefits are even more significant than in other types of structure. It has been said that the worst place to construct a building is on a building site and it is worth bearing this in mind when considering these points in more detail.

Off-site production
On-site activities are complicated by a great variety of factors, such as:

- Bad weather – not just ice and snow but, for example, concreting in hot weather creates difficulties.
- Overlapping trades and large quantities of labour in a relatively confined space plus the logistical problems of accommodating and providing for them.
- It is much easier to cast a concrete component on a factory floor than several metres above ground on a building site.

Control of quality
Any self-respecting precast concrete production facility will have a quality control system which will be an established routine and will probably be subject to third party surveillance. It is considerably more difficult to achieve a similar level of quality control under site conditions and at best there is an inevitable learning curve associated with each new site.

Speed of erection
One of the principal advantages of precast concrete is its speed of erection compared with site-intensive processes. Programme is always a driving force with any construction project whether the motivation is rental availability, the first match of the season, the turn of the millennium or some other crucial milestone.

The key feature of precasting is that in effect the superstructure is progressing off site while, or even before, the foundation work is being carried out.

Immediate working platform
When erecting a precast structure, its very nature usually means that an immediate working platform is provided. Floors can be immediately stood upon and worked from, staircases give immediate access to working levels and walls can provide an immediate enclosure to work areas. All of this has considerable advantages for following trades and offers significant benefits to the construction programme.

Minimum of temporary propping
Most precast components are designed on the basis of requiring a minimum of propping or even none at all. Comparison with insitu construction speaks for itself where often a

large number of props is required to support temporary shuttering. At best this is expensive equipment which is time-consuming to erect and to strike and at worst is impracticable because of high floor to ceiling dimensions.

Minimum of wet trades
Since the object of precasting is to cast the concrete off-site, it of course, follows that there is a major reduction or virtual elimination in the volume of concrete to be batched at site or delivered by other means. Precast walls as a replacement for conventional masonry avoids the need for large quantities of mortar, bricklaying labour, plasterwork and, again, the common factor of time.

All of these benefits can be applied with success to stadia construction.

Before discussing some of the typical precast units which have been used in almost every stadia built in the past decade, it is worth touching on some of the more adventurous projects which have been carried out in recent times.

STADIA BUILT USING PRECAST CONCRETE

The famous Saddledome at Calgary is a project where precast concrete was used to great effect. Precast concrete played a key role in this successful project through its use in panels in the roof, fixed to tensioned cables (Fig. 1). Calgary is an example of an all-purpose arena but precast concrete has been particularly successful for construction of grandstands for sport. Sandown Park and Goodwood are well-known British horse racing venues. Abu Dhabi in the United Arab Emirates is better known for camel racing.

All of these feature precast concrete cantilevered roof structures with a barrel vault

Fig. 1. The Saddledome, Calgary, Canada.

design comprising a primary beam of prestressed, precast concrete with an infill element of lightweight and often transluscent material (Fig. 2). Other major structural elements such as columns are also manufactured as precast concrete elements.

Watford Football Club (Figs 3 and 4) is another excellent example of the innovative and structurally exciting use of precast concrete. Whilst this stadium does not perhaps have the cachet of Calgary, Sandown Park or Goodwood, Watford are currently leading the Second Division of the Football League so – who knows ?

Fig. 2. Grandstand at Goodwood Racecourse, Sussex.

Fig. 3. Precast concrete frame construction, Watford Football Club.

Fig. 4. Watford Football Club.

STRUCTURAL ASPECTS

The general structural principles of stadia design are illustrated in Figs 5 and 6.

Fig. 5 shows a typical framing arrangement showing a semi-rigid frame for stability with columns jointed at mid-storey height and lateral bracing or framing as required. Fig. 6 shows a simple cantilever roof structure. Generally, this will suit smaller stadia such as are common in racing but the roof will probably dictate the overall structural design philosophy for more complex larger, football stadia and those designed for multi-purposes.

The recently completed stadium for Bolton Wanderers Football Club is a good example of innovative roof design which is structurally efficient and aesthetically pleasing (Figs 7 and 8).

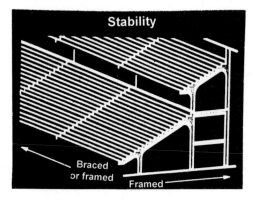

Fig. 5. Typical framed structure for grandstand terrace.

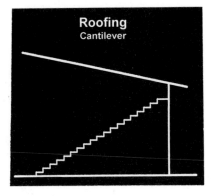

Fig. 6. Simple cantilever roof design.

Fig. 7. Bolton Wanderers Football Club. Main roof truss under construction showing temporary propping from precast concrete terrace.

Fig. 8. Bolton Wanderers Football Club. Completed corner tower supporting floodlighting and main structural trusses.

Most major grandstands built recently in the UK have been steel-framed with precast elements forming the minor structural elements. Although the term 'minor' is used in the context of overall structural design, the members are major in terms of volume and their effect on the economics and practicability of constructing a stadium.

Typical units which can be placed in this category are shown below. The large numbers and volumes of concrete involved in these types of unit for the Cardiff Millennium Stadium are shown by way of example.

Floors	6040 No.	50,000 m^2
Terrace units	5750 No.	6,000 m^3
Vomitories	74 No.	400 m^3
Staircases	18 No.	400 m^3
Steps	2500 No.	295 m^3

On the Cardiff Stadium there are also other units such as beams, rakers and walls all of which are precast concrete.

MANUFACTURE AND APPLICATION

This section expands briefly on the elements mentioned in terms of their method of manufacture, their application and some of their specific features.

Floors

Typically these are hollowcore slabs designed with or without a structural topping. Their method of manufacture is on a long line principle, extruded or slipformed on a heated steel bed with a length usually between 100 and 200 m.

Units are of course hollow and prestressed and therefore are extremely structurally efficient with regard to load/span performance. Grandstands are often on a 7 – 8 m grid and a 200 mm deep slab copes happily with this, combined with the loadings applied to a grandstand structure.

It is worth noting that the manufacturing process does not use a mould in the conventional sense – a fact which is not always known by designers and therefore on occasions results in impractical details being proposed. (Fig. 9.)

Some 6 million m^2 of precast floors are produced per annum in the UK for a wide variety of structures, including stadia, which bears testimony to the economic and practical advantages which they provide.

Terrace units

These are the most common unit with specific application to stadia and one therefore where careful attention to detail will pay worthwhile dividends.

Fig. 10 shows a typical terrace before the seating is installed. Once the seating is installed the casual user will probably not even notice the terrace units but the details are well worth a close look.

Bearing detail

Figs 11 and 12 show a typical seating on to a steel cleat attached to a raker beam. Note that

Fig. 9. Typical production of extruded concrete flooring.

Fig. 10. Typical precast concrete terrace unit layout prior to fixing of seating.

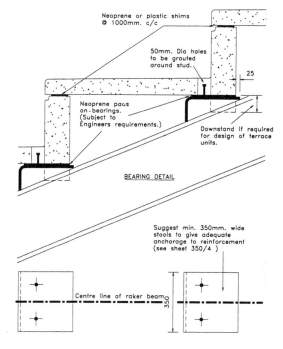

Neoprene or plastic shims
@ 1000mm. c/c

50mm. Dia holes
to be grouted
around stud.

25

Neoprene pads
on-bearings.
(Subject to
Engineers requirements.)

Downstand if required
for design of terrace
units.

BEARING DETAIL

Suggest min. 350mm. wide
stools to give adequate
anchorage to reinforcement
(see sheet 350/4)

Centre line of raker beam

350

PLAN OF BEARING STOOLS ON RAKER

Fig. 11. Typical bearing detail of precast concrete unit onto steel raker beam.

Fig. 12. Typical arrangement at front edge of beam.

the fixing is very simple – a hole in the terrace unit which engages a stud fixed to the cleat. This generally provides sufficient mechanical connection to ensure that the terrace deck acts as a diaphragm to the main structure.

All manner of alternative fixings are sometimes proposed, based on cast-in bolts, sockets, channels and so on. Apart from being very expensive in material cost, the unnecessary complication can make life very difficult on site when installing the units.

It is important to provide a seating cleat of adequate width.

Consideration of the reinforcement arrangement at the end of a reinforced concrete terrace unit taking account of cover, bend radii, anchorage lengths etc. will generally lead to a stool width of not less than 350 mm. See Fig. 13.

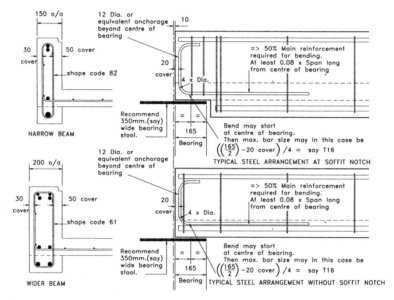

Fig. 13. Typical reinforcement arrangement at bearing of concrete terrace unit.

Stools of much narrower width are sometimes proposed simply because the reinforcement arrangement has not been properly considered at the appropriate time.

The actual width, depth and reinforcement content of the beam element of the terrace unit needs careful consideration. It is normally no problem to increase the depth of the beam, regardless of the riser height, by notching the end of the beam at the support in order to achieve the optimum of concrete and reinforcement content.

A modest adjustment to concrete section of 100 mm on depth reduces the reinforcement by two thirds without necessarily any increase in concrete area. This can result in savings of several tons of reinforcement on a typical stadium. When considering hundreds, or even thousands, of units, this kind of attention to detail is vital to the economics of a project.

Most terrace units are of reinforced concrete although for some projects they have been manufactured as prestressed concrete, but this can result in additional costs for special moulds and prestressing facilities.

One method of adopting a standardised approach using standard moulds but with the ability to provide the wide variation of risers and goings which are inevitable for different stadia and indeed within a single grandstand is to use an engineered adjustable mould. The unit is made 'toe down' and the is designed to be adjustable such that the going, the rise, the beam depth and width can all be varied. It is also constructed so that, with the exception of the back of the beam, all surfaces are cast against a steel shutter and thus a good surface finish is achieved on all seen faces.

Fig. 14 shows a typical terrace unit together with associated step blocks which can be fixed using an epoxy based material effectively 'gluing' the unit into place.

Fig. 14. Precast step units fixed to terrace units.

Vomitories, walls and staircases

Vomitory construction is another example where sensible rationalisation can produce a very effective solution using precast concrete walls and stair units.

Walls are normally a triangular or trapezoidal shape and are manufactured in a simple flat cast mould on a steel bed where features such as handrail pockets and terrace support fixings can be easily accommodated. (Fig. 15.)

Staircases in stadia vary from comparatively small elements associated with vomitories to very large units in major concourse areas. In all cases, these can be effectively provided in precast concrete so that a minimum of finishing is required after installation. (Fig. 16.)

Fig. 15. Simple vomitory

Fig. 16. Typical precast concrete staircase units.

SUMMARY

This paper is a brief overview of the basic types of precast units commonly used in stadia. From a manufacturer's view point I offer the following points in summary to designers.

- Keep it simple.
- Do not use complicated details unnecessarily
- Consider realistic tolerances
- Rationalise to achieve repetition
- Be aware of worst cases

An example of the last point: if a stadium has a standard grid of 7.5 m, in a corner, the units on the outside of the curve can be 10–12 m, which has many consequent implications for moulds, weight, craneage etc.

Earlier, the principal benefits of precast concrete were listed. These are repeated now, and it is hoped that by the examples and photographs shown, these advantages have been demonstrated to apply particularly to stadium construction:

- Off-site production
- Control of quality
- Speed of erection
- Immediate working platform
- Minimum of propping
- Minimum of wet trades

It is suggested that it is difficult to envisage building a stadium at the present time without using significant amounts of precast concrete in one form or another.

Acknowledgements
Thanks are due to Jan Bobrowski and Partners, consulting engineers for the stadia at Calgary, Goodwood and Watford; and to Tarmac Precast Concrete who manufactured the units at Watford and Goodwood.

12 MODERN STADIA HISTORY AND THE USE OF CONCRETE

M. A. LISCHER
HOK Sports Facilities Group, London, UK

SUMMARY: There is a growing demand for new and renovated stadia and an increasing market opportunity for the concrete industry. This paper traces the history of the modern stadia and explores the growing use of concrete.
Keywords: Air-supported structures, building types, concrete structures, history, moveable roofs, steel structures, tensegrity.

INTRODUCTION

The HOK Sports Facilities Group is the largest architectural practice in the world whose efforts are devoted exclusively to the design of new and renovation of existing stadia and indoor multi-purpose arenas. Since its founding in 1983, HOK Sport has been responsible for the design of 350 sports projects seating more than three million spectators. Projects currently under contract include the renovation of the Toulouse Stadium in France, in preparation for this summer's World Cup, design consultant to the Sydney 2000 Olympic Stadium and Arena, and renovation of Lancaster Park in New Zealand. Previous projects designed and constructed in the UK include: indoor arenas in Sheffield and Birmingham, the design of Manchester's Olympic Stadium used for the 2000 Olympic and 2002 Commonwealth Games Bids and the National Cycling Centre, an indoor velodrome also in Manchester. Needless to say our projects contain large amounts of concrete, both insitu and precast. The volume of the concrete in the 42,000 seat Jacobs Field Stadium in Cleveland, Ohio is 50,000 cubic metres.

Before discussing the concrete industry's potential market in sports facilities, it helps to put the modern stadium into perspective by reviewing the history of stadia.

THE STADIUM BUILDING TYPE

What we as designers do is nothing new. The Romans were building stadia 2,000 years ago. The Coliseum in Rome held 55,000 spectators and even boasted such modern innovations as a moveable roof. Although various cultures have constructed stadia through the ages, this analysis starts with what may be considered to be the birth of the modern stadium building type. In the author's opinion, this started in the early 1960s and is exemplified by the

Stadia, Arenas and Grandstands, edited by P.D. Thompson, J.J.A. Tolloczko and J.N. Clarke.
Published in 1998 by E & FN Spon, 11 New Fetter Lane, London EC4P 4EE, UK.   ISBN: 0 419 24040 3

Houston Astrodome's opening in 1965. This was the first fully enclosed multi-purpose stadium and began a tremendous period of stadium construction in North America and worldwide that continues today. There are social and cultural reasons why this is the start of modern stadia, including issues such as the development of television and the broadcast of live sports events to a growing worldwide audience. However, the technological development of modern stadia may be traced through long span roof design. The buildings that cultures through the ages construct represent the height of their technological prowess and leave a legacy for future generations to study. Next to religious architecture, no building type represents this better than stadia. If you think of a Roman building, you think of the Coliseum.

Steel

The roof structure of the Astrodome is constructed of steel, spans a circular diameter of 196 m and fully covers 60,000 spectators. Clearly this type of roof structure, rigid steel, was pushing the limits of the materials and the engineering technology of the day. Unfortunately, the volumes enclosed by these rigid steel domes were not big enough. The race for space was truly on in the 1960s, not only in the skies, but also on the ground. Roof technology had to catch up with the client's demands.

Air

The next breakthrough in long span technology was the development of air-supported structures. The concept is that of a balloon. Air is pumped into the building and the higher pressure indoor inflates a lightweight fabric roof restrained by cables. The first major stadium to use this concept was the Pontiac Silverdome, opened in 1975 near Detroit, Michigan. This seated 80,000 spectators in a 185 m by 235 m footprint area. Using air to support the huge spans of a roof was a great idea and the change to lightweight fabric enclosure materials was fundamental to the increase in spans. Although the last major air supported dome opened in 1983 in Indianapolis, Indiana, this type of structural system was outdated even before the Silverdome opened its doors. One major event in history sounded the death knell for large air-supported structures virtually overnight. The Arab oil embargo of 1973 and the rocketing cost of electricity increased the cost of keeping 'the roof up' dramatically. Large amounts of electricity are required to keep the fans running 24 hours a day, 365 days a year, to pressurise the interior space of these buildings.

The engineers had to get back to work and develop a structural system that was still very light in weight yet could maintain its shape and structural integrity without a continuous drain on the facility's revenue streams.

Tensegrity

The air-supported roof structures evolved into the tensegrity structure or 'cablenet' dome. The first major manifestation of this structure was in Seoul, Korea for the indoor arena constructed for the 1986 Korean Olympic Games. The system is fascinating and involves exploiting the properties of tensile and compressive forces. All tensile forces are resisted by strands of cable and the compression forces by slender steel columns. The mathematics to design this system are so complicated, that patented computer software is required that works on a non-linear analysis principle. The most recent and largest cablenet stadium structure is the Georgiadome in Atlanta. Opened in 1992, it seats 71,000 spectators. It was featured on worldwide TV in 1996 as it housed the gymnastics competition at the Atlanta Olympics.

Moveable roofs

Our clients continue to demand more and they now want the ability to have an enclosed stadium to protect events from inclement weather and the ability to hold events in the open air. Design has gone full circle back to the Coliseum, as the next phase in stadium development is the implementation of moveable roofs. The first working example is the Toronto Skydome in Canada opened in 1990 at a cost of some $500 million. The Fukuoka Dome in Japan followed in 1994. The problem with these two facilities is the engineering of the roof has dominated and compromised the design of the stadium below. The roof structures dictated circular seating bowls which compromise the sight lines and spectator proximity for most events.

European stadia are particularly suited to the use of moveable roofs since FIFA requires football to be played in an 'open-air' stadium. The first to be constructed is the Amsterdam Stadium with an opening of approximately 7,900 square metres. Stadia with moveable roofs are also under construction in Arnhem, Holland and Cardiff, Wales. A number of others are currently under consideration or design.

The latest generation of moveable roof stadia has just begun construction in the United States. These roofs open completely exposing the entire seating bowl area to the outdoor environment. The first couple of facilities are baseball only stadia with the Bank One Stadium in Phoenix, Arizona, scheduled for completion this year. This roof contains 20,000 tons of steel and is articulated much like a telescope. It can be opened with the use of electric motors in approximately six hours. Another, soon to begin construction, is a new home for the Milwaukee Brewers Baseball Team in Milwaukee, Wisconsin. This moveable roof is called a fan dome and opens much like a Japanese fan.

This brings us up to date with the evolution of stadia design. Of importance to your industry is that the modern stadium with or without a moveable roof contains more concrete than ever before in history.

USE OF CONCRETE

Because of its appearance and low maintenance cost, concrete will continue to play a major role in the construction of most new stadia. However, its massive appearance and heavy characteristics that designers often desire sometimes works against its use.

Where large sections of a stadium move and articulate, steel is generally used. For a given span or structural bay, a steel structure is much lighter and better able to accommodate the dynamic loads applied while moving. In regard to design aesthetics, concrete's low maintenance is sometimes traded off for steel's light and delicate appearance.

Stadia are massive structures and, when desired, designers can play certain visual tricks in order to reduce the imposing appearance of a given stadium. A couple of examples illustrate this point. At Oriole Park in Baltimore, Maryland, the new stadium was constructed within an existing city neighbourhood. In order to minimise the stadium's impact on the scale of the neighbourhood, the pitch was constructed 12 m below the existing street level. In addition, the structure for the upper seating decked is stepped back from the brick clad facade and constructed of dark painted steel. When viewed from street level, the brick element is of a similar scale, colour and material to the surrounding buildings. The steel upper deck simply recedes from view and the dark colour is very unobtrusive when seen.

In Washington DC, the new Jack Kent Cooke Stadium was constructed in a 'green field' site surrounded by landscaped car parking. The sculptural qualities of the stand-alone building were considered when selecting materials. The design uses insitu concrete for the external ramps and superstructure for the lower two-thirds of the building. This creates a visual base for the stadium and anchors the mass to its site. Above the upper concourse, the structural columns splay into a lightweight web of structural steel. The structure soars above the seating bowl and terminates as the structure carries the field lighting system.

CONCLUSION

Concrete will continue to play a major role in the design of new stadia and most designers and specifiers are enthusiastic about its appearance. However, as increasingly intricate roof structures and more sophisticated design aesthetics are required, steel and other materials will challenge concrete's use. The concrete industry and architectural profession must maintain an open dialogue to explore new and interesting applications for concrete and designers must maintain their awareness of concrete's improving technology and capabilities. Events such as Stadia 2000 are a perfect forum for this interaction. Only then can designers take full advantage of concrete's potential and the concrete industry fully exploit a growing market.

13 STEEL STANDS: THE PREMIER CHOICE FOR SPORTS STADIA

P. J. BISSEKER
Technical Marketing, British Steel plc, UK

SUMMARY: The paper reviews the benefits of using steel for constructing sports stadia, and describes particular aspects where steel construction enables best use to be made of the facility. These include the roof structure, and under terrace structure. Speed of construction is a special advantage of steel. Examples are given of recent stadia projects where the architectural expression of the structure has resulted in exciting and innovative structural solutions.
Keywords: Column-free areas, future developments, roofs, speed of construction, steel structures, under terrace structures.

INTRODUCTION

The versatility of structural steelwork, enabling architects and engineers to give full vent to safe, innovative, attractive and varied design concepts, has led a revolution in UK sports stadia, and catering for the demands of spectators well into the next millennium.

Following the Hillsborough disaster, Lord Justice Taylor's report (1990) outlined clear mandatory safety requirements for sports stadia incorporating all aspects of crowd safety. This was the catalyst and, since 1991, 164 new stands have been built (counting new stadia as four stands) incorporating an estimated 100,000 tomes of structural steelwork and steel cladding.

The difficult task facing stadia and sports ground owners is to provide the safe, comfortable and unobstructed viewing environment deserved and demanded by loyal supporters while ensuring that they get the best value for money from the construction or reconstruction involved in ground improvements.

ROOF STRUCTURE

The first structural consideration is to provide an uninterrupted view of the playing field and the flexibility of steelwork gives the architect/engineer a number of choices:

Traditional cantilever arrangement
The cantilever is supported from the rear of the stand and is suitable for the smaller club (Fig. 1) which may only wish to cover 8 or 10 terrace steps, up to the magnificent, recently

Stadia, Arenas and Grandstands, edited by P.D. Thompson, J.J.A. Tolloczko and J.N. Clarke.
Published in 1998 by E & FN Spon, 11 New Fetter Lane, London EC4P 4EE, UK.   ISBN: 0 419 24040 3

Fig. 1. Crawley Athletic Stadium.

built Manchester United North Stand, which boasts a cantilever roof of 58 m, covers three tiers of seating and increases the all-seater capacity of Old Trafford to 55,000.

Fig. 2. Manchester United North Stand (completed for 1996/97 season).

The new Shanghai Stadium has a massive tubular steel cantilever roof of 73 m extending round the whole stadium. At Newcastle United, a 65 m steel tubular cantilever roof to incorporate a further tier of seating to the North and Western Stands is proposed in a planned upgrading of St James Park.

Strength to weight ratio

The high strength to weight ratio of structural hollow sections (SHS) provides an economic solution for spanning the full length of a soccer pitch (approximately 100 m). This arrangement was first adopted in preference to the original design incorporating two columns and three girders as far back as 1970 when Bristol City built the Dolman Stand to seat 5,400 spectators. Severe access problems existed for craneage at the rear of the stand, quite a common problem with many stadia, and this was overcome by the design of a light tubular main girder weighing around 60 tonnes. The girder, with chord ends bevelled in preparation for site welding, was fabricated in five sections, transported on a low loader to site, and lined up on stillages. The five sections were then joined by in-line full-strength butt welds on site (main chords 457 mm diameter, maximum thickness 28 mm, circular hollow sections (CHS)) and the whole girder lifted into place on UC end columns in 20 minutes. The remainder of the roof steelwork comprised simple CHS trusses spanning from the main girder to simple UC columns at the rear of the stand and simple CHS cantilever girders between the main girder and the front edge of the stand (see Figs 3 and 4).

This was quickly followed by a similar arrangement at Celtic Park Glasgow where a twin cross-braced main tubular girder spanning approximately 100 m facilitated a clear uninterrupted view for 10,000 seated spectators.

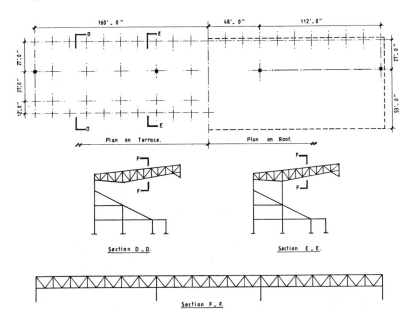

Fig. 3. Plan and sections of the Dolman Stand, Bristol City, as originally conceived.
The arrangement of the area below the terracing was unsatisfactory as it was difficult to use the entire area available at first floor level. Also, the relatively small column grid was not suited to bowling greens.

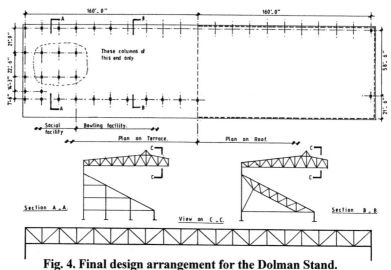

Fig. 4. Final design arrangement for the Dolman Stand.
The social and bowling green facilities are concentrated at ground level, which involves less cost.
The 320 ft-long, 15 ft 6 in-deep, main girder has main chords of 18 in circular hollow sections.

In the current post-Taylor era a number of stands have adopted this format. Perhaps the most spectacular and certainly the largest span is the triangular CHS girder at Ibrox Park, Glasgow, which spans 145 m and enabled Glasgow Rangers to add a further tier to its

Fig. 5. Ibrox Park Glasgow. Triangular CHS roof girder, 145 m span. Three-tier steel terrace structure – additional tier increased stand capacity by 4000 seats.

existing stand while retaining its coveted 'listed' front buildings housing boardroom, trophy room and other important facilities.

Curved arches
Curved arches provide a particularly architecturally pleasing arrangement and these have recently been featured in stadia at Huddersfield (Alfred McAlpine Stadium) and in Hong Kong (240 m span), Figs 6 and 7. The simple trusses carrying the roof can easily be clad with traditional steel sheeting, a wide range of profiles and colours are currently available, or fabric. Two major stadia – the Olympic Stadium in Sydney and the New Millennium Stadium, Cardiff – feature longitudinally spanning tubular arches.

Fig. 6. Alfred McAlpine Stadium, Huddersfield. Three triangular CHS arches.

Fig. 7. Hong Kong Stadium for the Royal Hong Kong Jockey Club.
40,000 seater stadium incorporating two 240 m-span tubular arches.
28,000 seats completed in time for the 1993 Rugby Sevens competition.

UNDER TERRACE STRUCTURE

A great advantage of building in steel is the ease with which wide open spaces can be created. This is particularly important in stadia since more flexibility in the area beneath the terracing, gives more scope to the owner to provide income-generating facilities.

The Dolman Stand at Ashton Gate, Bristol City is one of the earliest examples of this concept. The final scheme eliminated a forest of columns from the original design, providing a column-free area which housed indoor bowling, a popular leisure past-time in late 1960s/early 1970s. It also allowed flexibility for many other uses to meet future changing needs of the public.

Further significant examples of open space under terrace steel design are found at Villa Park, Birmingham (Fig. 8) where a nominal 25 m-wide concourse has been created for fans to meet, congregate and enjoy refreshments, and at Old Trafford, which accommodates United Road in a nominal 16.6 m span aisle.

Fig. 8. Villa Park Birmingham.
Two-tier stand with under terrace steelwork providing massive 25 m-wide spectator concourse.

SPEED OF CONSTRUCTION

Spectators are a valuable source of income and so speed of construction is of prime importance, minimising the period from when existing terraces are taken out of use until the new facilities are opened to the paying spectators. With careful planning, most of the construction can be planned for the off-season, keeping inconvenience to the Club and fans to a minimum. Steel is the obvious choice. The steelwork can be fabricated accurately within a high QA regime at the fabricator's works towards the end of a soccer season and quickly assembled on site during the off season so that following trades can complete the stand often in time for the start of the next season. Again one of the most significant recent examples of rapid and convenient construction is the new North Stand at Manchester United where demolition of the existing stand was started in May 1995 and the new stand opened in June 196 in time for Euro 96. Not only was the all-seater ground capacity increased by 22,000 in just 11 months but the lower deck of seats had been available for a significant part of the 1995/96 season.

UNIQUE STRUCTURE

The most spectacular addition to the UK scene is the new Reebok Stadium for Bolton Wanderers (Fig. 9), with steelwork designed, fabricated and erected by Watson Steel Limited in time for the 1997/98 season. The dominant features of this green field site stadium are four curved trusses inclined towards the pitch (two with 150 m span and two with 110 m span) supported on four 50 m-tall three-legged steel pylons at the corners of the stadium. The extended pylons also carry the floodlights.

Fig. 9. The unique Reebok Stadium for Bolton Wanderers FC.

THE FUTURE

In the long-term programme, a further 144 projects have been identified including major national stadia. Sliding roofs have been discussed for many years and the first of these is being incorporated in the New Millennium Stadium, Cardiff for the Welsh Rugby Union (under construction now and due for completion for the 1999 World Cup). Further sliding roof stadia are being planned. There has also been a proposal for a stadium in which the pitch, in sections, is moved horizontally under the terraces, and even for a stadium where the pitch can be jacked up to provide a roof cover over other sporting facilities during the period between soccer matches.

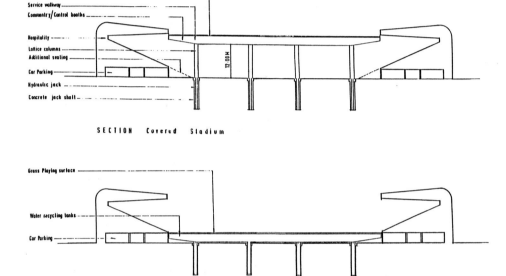

Fig. 10. Artist's impression of futurist stadium by RENA with raised pitch providing cover for other sports activities.

CONCLUSION

The current atmosphere in stadia worldwide, backed by the impressive record of stadia construction in the past decade, emphasises that steel not only stands for safety, innovation, flexibility and strength with elegance, it is the premier choice in value for money.

14 FULL-SCALE TESTING OF LARGE CANTILEVER GRANDSTANDS TO DETERMINE THEIR DYNAMIC RESPONSE

J. D. LITTLER
Building Research Establishment Ltd, Garston, UK

SUMMARY: Full-scale tests have been conducted on four large cantilever grandstands to determine their dynamic response. Tests on the empty stands showed they had fundamental vertical natural frequencies which ranged from 4.6 Hz to 6.8 Hz. Each stand was also monitored during a pop or rock concert. Perceptible excitation of the stands by human action occurred at all of the concerts, almost all of the response occurring at the beat frequency of the music being played or one of its harmonics. Maximum measured values of single peak acceleration ranged from 0.48 m/s² to 1.62 m/s² (4.9 to 16.5% g). The paper gives details of the measurements made on the stands both when empty and during the concerts, gives examples of the results obtained, and discusses the safety of adopting lower frequency criteria for large cantilever grandstands than those set out in BS6399: Part 1.
Keywords: cantilever grandstands, code of practice, natural frequency, serviceability, synchronised dynamic loading.

INTRODUCTION

When it was introduced in September 1996, BS 6399: Part 1 [1] became the first British Standard which included a section on synchronised dance loading. It stated that any structures which might be subjected to this loading should be designed in one of two ways. Either by designing the structure to have natural frequencies above specific threshold values in which case the problem was avoided; or by ensuring that the structure was designed to be able to withstand this type of loading. The threshold values of natural frequency given in the standard are 8.4 Hz in the vertical direction and 4.0 Hz in the horizontal direction. Many large cantilever grandstands are being designed and built at present in the UK, but it is difficult for them to be both designed economically and to meet the frequency thresholds. Consequently, there has been considerable debate as to whether the frequencies given in BS6399 can be safely reduced in the specific case of cantilever grandstands, and whether the synchronised loading case in BS6399 was appropriate for sporting events as well as pop and rock concerts.

Over the last two years the Building Research Establishment Ltd (BRE) has carried out full-scale tests at three stadia in different countries of Western Europe to determine the dynamic response of their large cantilever grandstands. Two sets of test have been performed on each tier of each stand. Firstly the empty stand has been subjected to forced

Stadia, Arenas and Grandstands, edited by P.D. Thompson, J.J.A. Tolloczko and J.N. Clarke.
Published in 1998 by E & FN Spon, 11 New Fetter Lane, London EC4P 4EE, UK.   ISBN: 0 419 24040 3

vibration testing to determine the dynamic characteristics (natural frequency, damping, mode shape) of its lowest frequency (fundamental) vertical mode. Secondly the response of the stand was measured during a rock or pop concert.

BACKGROUND

The frequency thresholds of 8.4 Hz and 4 Hz given in BS6399 [1] are for the relevant modes of vibration of an empty structure. It is important to understand the reasons why these frequencies are given for an empty structure but for a loading condition which involves a full stand.

Previous BRE and other studies have shown that the most severe form of human loading for these structures is likely to be when people at a pop or rock concert jump up and down to the beat frequency of the music. This has been shown in analytical studies [2] as well as laboratory based [3] and field measurements [4]. When people jump like this they produce responses at the beat frequency of the music as well as at the integer multiples of that frequency (the harmonics). It is only by summing the response at each of these frequencies that the total response can be calculated [5]. In practice, taking the response in the first three harmonics is usually considered to be sufficient for an evaluation of safety1. If these loads excite a natural frequency of the structure then resonance will occur which can greatly amplify the structural response. As mentioned above, BS6399 provides two ways of coping with the synchronised dance loads that occur at pop and rock concerts. The first is to design the structure so that its natural frequencies do not coincide with the possible input frequency of the people jumping or its second or third harmonic. The second is to design the structure to withstand this loading.

Unless the structure is to be specifically designed to cope with these synchronised dance loads, its natural frequencies must be above the third harmonic of the dance frequency of the input loading. BS6399 considers that for large groups the frequency range for this type of loading should be 1.5 Hz to 2.8 Hz as co-ordinated movement at frequencies higher than this is impractical. The 8.4 Hz given in BS6399 is derived from this frequency being three times 2.8 Hz. BS6399 also specifies that the frequencies should be evaluated for the appropriate mode of vibration of the empty structure. This is despite that fact that the frequency of a grandstand with a crowd on it will not be the same as the empty grandstand. However, the frequency of an empty stand is used for a number of important reasons:

- The worst loading case is when all the people on the stand jump up and down together. In this scenario, the people are in the air for the part of the loading cycle just prior to the imposition of the main load, they are not interacting with the stand, and consequently the natural frequency of the stand when empty is the one to be considered.
- A structure can be tested on completion before being used by the public to check whether the frequency criteria are satisfied. This is also true where an existing structure is to be assessed for use during a pop or rock concert.
- It is extremely difficult if not impossible to predict the dynamic characteristics of the full stand as these depend on the combined human-structure system. The people on the stand cannot be adequately represented by a simple additional mass [6].

One example of this latter phenomenon can be illustrated by a test carried out by the author on a retractable stand with 99 seats. The fundamental front-to-back mode of the empty stand was 3.05 Hz. An identical test was then conducted on the stand but with people sitting in all the seats. A frequency of 1.71 Hz was obtained. Finally all the people on the stand were asked to stand up and this time a frequency of 3.30 Hz was obtained. Damping also varied considerably between the three tests. Whilst this example is for the horizontal mode of a retractable stand, it does illustrate that the people on a stand cannot be adequately represented by a simple additional mass.

If a structure cannot meet the frequency thresholds given in BS6399 but is to be used at events where synchronised dance loading may take place, then clearly the structure must be designed to withstand these loads.

DETERMINING THE CHARACTERISTICS OF THE EMPTY STANDS

The BRE tests on the lower tier of a three-tiered stand (stand A) will be used as an example of the testing procedure. Only the testing in the vertical direction is described, although tests in the horizontal directions were also carried out on some stands. Firstly, the response of an accelerometer to ambient (wind) excitation was recorded. This was then processed using a fast Fourier transform (FFT) to produce an autospectrum from which the modal frequencies can be identified (Fig. 1). Impact tests, where the response of the stand is measured as someone stood on the balls of his feet before letting his heels drop, were then carried out. The autospectrum from one of these tests is shown in Fig. 2. Both test methods indicate modes at about 4.8, 6.0 and 7.3 Hz. The fundamental mode was then investigated further by strapping an eccentric mass exciter to the stand. Fig. 3 shows the response spectrum obtained as the frequency of the exciter was incremented through the frequency range of interest and the response measured by an accelerometer positioned nearby. The best fit single degree of freedom curve to the relevant part of the experimental data is also shown. This best fit curve had values of 4.69 Hz and damping of 1.4% critical. As the force imparted during the steady-state tests is much larger than that in the impact and ambient tests, larger amplitude motion is induced and a lower natural frequency in the steady-state tests is expected because of the non-linear response of the stand. This is not unusual and is seen in many different types of structure. The damping in this mode was investigated further by setting the exciter to the frequency of maximum response and then monitoring the decay of oscillation. A theoretical decay was then fitted to part of the experimental decay (Fig. 4). This yielded a damping value of 1.32%. Whilst this is not a particularly good fit with the experimental data it confirms a damping value of about 1.3 to 1.4% critical. Finally, the exciter was restarted and set to the natural frequency. An accelerometer was then moved around the stand to measure the response (amplitude and phase) in various positions relative to that of a second accelerometer kept in a fixed position. The mode shape obtained showed that the whole tier was acting as a single body in this mode.

The other two tiers of the stand were tested in an identical manner. In the steady-state forced vibration tests the middle tier had a fundamental vertical natural frequency of 4.93 Hz with damping of about 1.6% whilst the equivalent figures for the upper tier were 5.70 Hz and 2.5 to 2.6% critical.

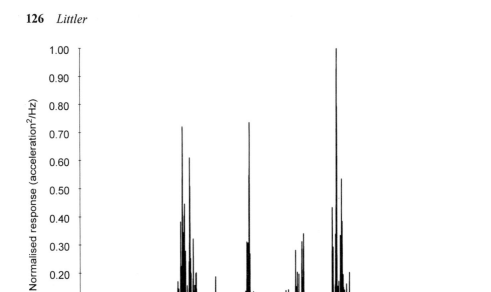

Fig. 1. Autospectrum from ambient monitoring in the vertical direction on lower tier of stand A.

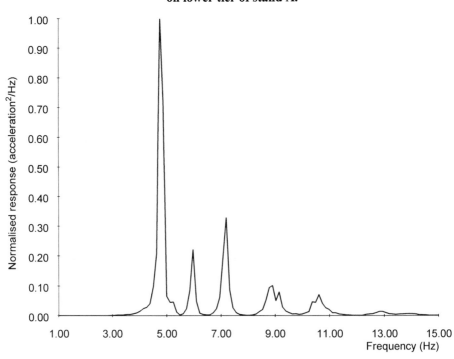

Fig. 2. Autospectrum from vertical impact test on lower tier of stand A.

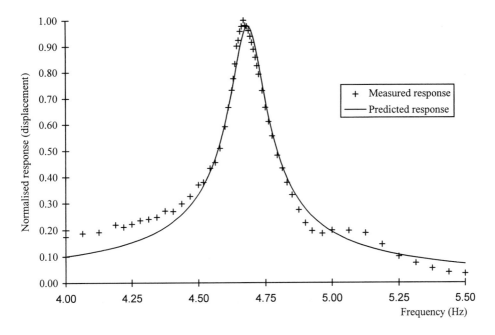

Fig. 3. Curve fitting around the fundamental vertical mode of the lower tier of stand A as identified in steady-state tests.

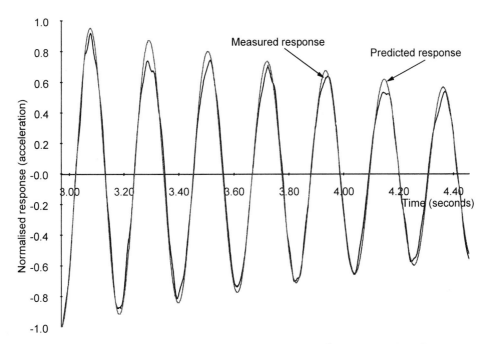

Fig. 4. Part of decay of oscillation in the fundamental vertical mode of the lower tier of stand A, with the best fit theoretical decay.

Impact tests only were carried out on stand B when it was empty. The upper and lower tiers had fundamental vertical natural frequencies of 4.64 Hz and 4.76 Hz respectively. Steady-state forced vibration tests were carried out on stands C and D which are stands of different design at the same stadium. The fundamental vertical natural frequencies obtained were 6.36 Hz on stand C and 6.79 Hz (with damping of about 4%) on stand D. It is interesting to note that stand D (which is over ten years old) had by far the highest damping value obtained whereas stands A and B (which are both less than five years old) had much lower damping values.

MONITORING THE STANDS DURING POP AND ROCK CONCERTS

After the tests on the empty stand were completed, each stand was monitored during at least one pop or rock concert. The testing procedure was the same in each case.

Accelerometers sometimes supplemented by geophones were mounted in one or more boxes underneath the front edge of the cantilever (or in two cases under a seat on the front row of the stand occupied by one of the BRE staff) and the signals from them fed to a digital tape recorder. The response from each transducer was analysed by dividing the recording into between 100 and 300 contiguous records, each of either 51.2 or 102.4 seconds duration. Each of these records was then analysed to obtain the peak acceleration or velocity and the frequency of maximum spectral response. On stand A (all three tiers) and on stand B the fundamental vertical frequency of the full stand was between 0.35 Hz and 0.5 Hz less than the frequency of the empty stand.

As stated above, the fundamental vertical natural frequency of the empty stands ranged from 4.6 Hz to 6.8 Hz. If the 2.8 Hz given in BS6399 is taken as the upper limit at which co-ordinated jumping can take place, then the natural frequencies on the tested stands can only be excited by the second or higher harmonic for those structures with frequencies between 4.6 Hz and 5.6 Hz, and by the third or higher harmonic for those structures with frequencies above 5.6 Hz. Motion of the stands was perceptible at all of the concerts, and in some cases this motion was clearly perceptible and would be 'disturbing' under at least one classification system [7]. Whilst several people remarked about the movement of two of the stands, as far as the author is aware, nobody complained about it or questioned the structural integrity of the stands. The maximum unfiltered single peak accelerations ranged from 0.48 m/s^2 to 1.62 m/s^2 (4.9% to 16.5% g). Almost all of the response occurred at the beat frequency of the music being played or one of its harmonics. All the recorded data for stands A and B were passed through a 10 Hz low-pass filter and the maximum single peak accelerations after filtering ranged from 0.20 m/s^2 to 0.80 m/s^2 (2.0% to 8.2% g). This shows that suggested serviceability criteria of 5% or 7% g are already being exceeded at many concerts and even if only accelerations filtered at 10 Hz are used, the suggested limits are still being reached, even when the audience behaviour is not particularly exuberant.

Fig. 5 shows the frequency of maximum spectral response under 6 Hz in the vertical direction on the lower tier of stand B. Fig. 5 was obtained during a concert by a rock band where the majority of the audience were 16 to 25-year-olds and about two-thirds male. A large number of the audience tended to jump up and down on the spot in time to the music ('pogoing') at the start of each song for a period of 20 to 30 seconds. Fig. 5, where each record is 51.2 seconds long, shows that the frequencies of people dancing to the beat of the

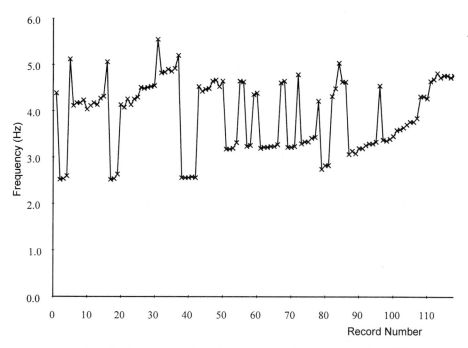

Fig. 5. Frequency of maximum spectral response under 6 Hz, stand B, vertical direction.

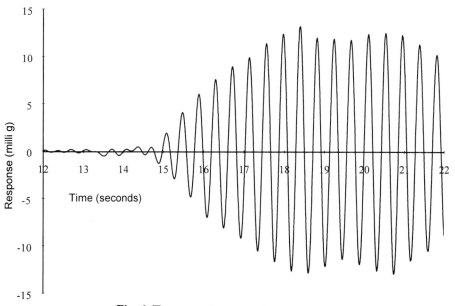

Fig. 6. Ten second extract from record 111 on stand B, vertical direction.

music are evident at most points throughout the recording. The beat frequencies of the songs played ranged from 1.33 Hz to 2.61 Hz. Fig. 6 is typical of what occurred at the start of each song during the concert on stand B. It shows a ten second extract from the filtered recording which features a very rapid increase in response at the start of a song with a beat frequency of 2.32 Hz.

Fig. 7 is the autospectrum for the whole of the unfiltered record from which Fig. 6 is an extract. This shows that the response is almost entirely at either the beat frequency of the music (2.32 Hz) or at the second harmonic (4.65 Hz) which is close to the fundamental frequency of the empty stand (4.76 Hz in impact tests). By filtering the data at 3 Hz low-pass (see Fig. 6) it is possible to obtain a response at a single frequency and hence calculate the equivalent displacement in the fundamental mode. However, it should be noted that the displacement obtained will be less than the actual displacement as it contains none of the response at the second harmonic, which, as Fig. 7 shows, is not insignificant. In this case the calculated displacement from the data filtered at 3 Hz low-pass was 1.25 mm peak to peak while that calculated from the data band-pass filtered between 3 and 6 Hz was 0.752 mm peak to peak. However, these two figures cannot be simply summed to give the overall displacement.

For stands A and B all the response data were filtered to remove the second and higher harmonics and then the equivalent displacement calculated assuming that all the response occurred at the beat frequency of the music. This yielded maximum peak to peak displacements on stand A of 2.9, 2.7 and 2.4 mm on the middle, lower and upper tier respectively and 1.35 mm on the lower tier of stand B. In each case the spectra for these occasions were checked to ensure that the assumption was valid. It should be noted that the maximum accelerations did not occur at the same time as the maximum calculated displacements. The latter events were associated with slightly lower levels of acceleration than the maxima but occurred during songs with slower beat frequencies.

The calculated displacements quoted above for stand A took place during a concert by an artist who has been popular for about 30 years and attracted an audience with a wide age range. The maximum accelerations and calculated displacements took place when the audience were standing and clapping in time to the music. There was not any widespread jumping up and down as occurred during the concert on stand B. One further feature of interest was that the audience were inducing a large response in the roof of the stand. This was very apparent because some speakers were attached to the cantilever tip of the roof and these were visibly moving backwards and forwards in time to the beat frequency of the music. Consequently, in some circumstances, roof motion can be excited by spectator activity [8].

Stand A was monitored during a second concert when the crowd were somewhat less animated than at the first concert possibly due to the inclement weather. However, the maximum calculated peak to peak vertical displacements for each tier varied from 0.92 mm to 2.6 mm. The maximum calculated peak to peak front-to-back displacements on the middle and lower tiers were 0.58 mm and 0.50 mm respectively. Figures 8 and 9 show two spectra obtained on the upper tier during this second concert. Fig. 8 is an autospectrum obtained during a song with a beat frequency of 1.97 Hz when all the maximum calculated vertical displacements occurred. This was the only song where the largest spectral response was at the beat frequency of the music. In all the other cases it was at the second harmonic (the harmonic closest to the natural frequency of the tier) and therefore similar in form to Fig. 7. Fig. 9 shows an autospectrum obtained during the song with the fastest beat

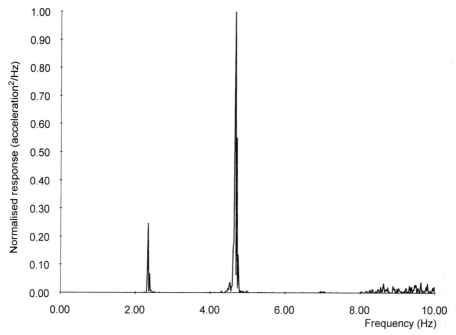

Fig. 7. Autospectrum from record 111 during song with beat frequency of 2.32 Hz, stand B, vertical direction.

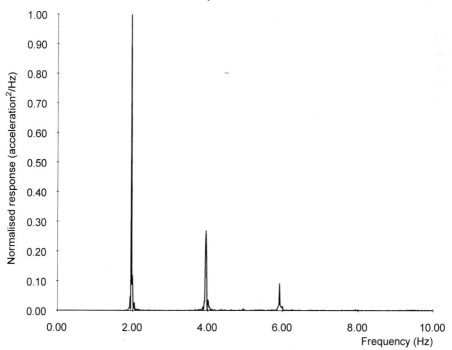

Fig. 8. Autospectrum during song with beat frequency of 1.97 Hz, stand A, upper tier, vertical direction.

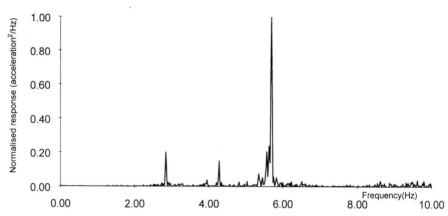

Fig. 9. Autospectrum during song with beat frequency of 2.84 Hz, stand A upper tier, vertical direction.

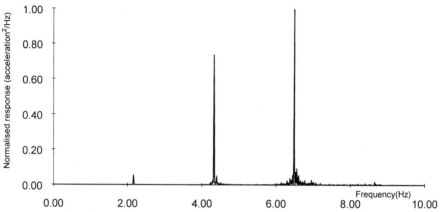

Fig. 10. Autospectrum during song with beat frequency of 2.17 Hz, stand D, vertical direction.

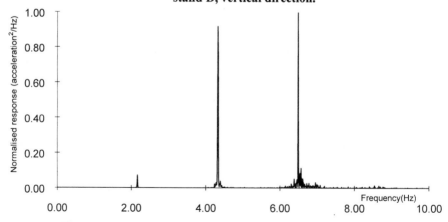

Fig. 11. Autospectrum during song with beat frequency of 2.17 Hz, stand D, front-to-back direction.

frequency (2.84 Hz). This shows a response at the beat frequency, the main response at the second harmonic, and also a response at 1.5 times the beat frequency. This could indicate that at this frequency (just beyond the maximum frequency for co-ordinated jumping given in BS6399) some of the audience were responding to the music but only on every other beat. Thus the response at 1.5 times the beat frequency could be the third harmonic of people moving at half the beat frequency of the music. However, there is no evidence of a peak in the autospectrum at half the beat frequency of the music.

Stands C and D were monitored during a rock concert by a band which attracted an audience whose ages varied from mid teens to late forties. However, this concert took place during a period of national mourning and the crowd were somewhat subdued throughout the concert. Figs 10 and 11 show autospectra obtained from stand D during a song with a beat frequency of 2.17 Hz. The largest accelerations on both stands C and D were measured during this song and motion of the stand was perceptible. Some, but by no means all the people on the stand were dancing on the spot and clapping in time to the music, but no 'pogoing' was seen. Fig. 10 shows the response in the vertical direction where the peak response was 0.48 m/s^2 while Fig. 11 shows the response in the front-to-back direction where the peak response was 0.25 m/s^2. Figs 10 and 11 demonstrate that even stands with natural frequencies in excess of 6 Hz can be excited by synchronised human actions. Fig. 11 also demonstrates that the horizontal as well as the vertical response has components at the beat frequency and its harmonics.

THE NEED FOR FURTHER RESEARCH

The latest edition of the Guide to Safety at Sports Grounds [9] calls for threshold values of 6 Hz in the vertical direction and 3 Hz in the horizontal direction even at events other than pop concerts. The paper by Reid, Dickie and Wright [8] has the same threshold values but suggests that they be applied to all loading cases. The author is not aware of any structural problems that have occurred with large cantilever grandstands having vertical natural frequencies between 6 Hz and 8.4 Hz, or horizontal frequencies between 3 and 4 Hz. However, the author is also not aware of any evidence which demonstrates that structures with natural frequencies within these ranges cannot have structural problems caused by synchro-nised dynamic loading. Indeed, the responses measured on stands C and D have shown that stands with vertical natural frequencies within this range can be excited at resonance by this type of loading, although clearly this does not imply that perceptibility equates to any kind of structural deficiency. It must be stressed that adopting the frequencies given in BS6399 does not preclude the design of stadia with lower natural frequencies. However, it does mean that where the designed frequencies are lower than the threshold ones, the stand must be shown to be able to withstand the type of loading that it may be subjected to. This is what would be required anyway under the Guide to Safety at Sports Ground criteria if a design had frequencies lower than 6 Hz and 3 Hz (even if the stand was not likely to be subjected to synchronised dance loading) although the way such a dynamic evaluation is carried out differs from that in BS6399.

There is little doubt that further research is required in this area to see whether the frequencies given in BS6399 can be safely reduced in the specific case of large cantilever grandstands. Two main areas for this further research are the co-ordination factors for large groups of people and the whole field of the dynamic response of these structures in the horizontal directions. Research in this first area needs to address how different stimuli effect synchronised actions. Whilst some of the lack of co-ordination evident with a large group of people can be attributed to the inevitable time lags that occur at large stadium concerts; the increasing use of large video screens means that the whole audience in a stadium can be synchronised without any time delay.

CONCLUSIONS

Several large cantilever grandstands have been tested. The fundamental vertical natural frequency of the empty stands ranged from 4.6 Hz to 6.8 Hz. Damping values ranged from 1.3 to 4% critical. Each stand was monitored during at least one pop or rock concert. Perceptible excitation of the stands by human action occurred at all of the concerts, almost all of the response occurring at the beat frequency of the music being played or one of its harmonics. Maximum measured values of single peak acceleration ranged from 0.48 m/s^2 to 1.62 m/s^2 (4.9% to 16.5% g).

REFERENCES

1. British Standards Institution. BS6399: *Loading for buildings* Part 1: 1996 *Code of practice for dead and imposed loads*. BSI, September 1996.
2. Ji, T and Ellis, B R. Floor vibration induced by dance-type loads: theory. *The Structural Engineer*, Volume 72, No. 3, 1 February 1994, pp 37–44.
3. Ellis, B R and Ji, T. Floor vibration induced by dance-type loads: verification. *The Structural Engineer*, Volume 72, No. 3, 1 February 1994, pp 45–50.
4. Littler, J D. Measuring the dynamic response of temporary grandstands. *Structural Dynamics – EURODYN'96*, Augusti, Borri & Spinelli (eds.), Balkema, Rotterdam, 1996, pp 907–13.
5. Building Research Establishment Ltd. Digest 426. *The response of structures to dynamic crowd loads*. Construction Research Communications Ltd, October 1997.
6. Ellis, B R and Ji, T. Human-structure interaction in vertical vibrations. *Proc. Instn Civ. Engrs Structs & Bldgs*, 1997, Volume 122, Feb., pp 1–9.
7. Kasperski, M. Actual problems with stand structures due to spectator-induced vibrations. *Structural Dynamics – EURODYN'96*, Augusti, Borri & Spinelli (eds.), Balkema, Rotterdam, 1996, pp 455–61.
8. Reid, W M, Dickie, J F and Wright, J. Stadium structures: are they excited? *The Structural Engineer*, Volume 75, No. 22, 18 November 1997, pp 383–8.
9. The Scottish Office and Department of National Heritage. *Guide to safety at sports grounds* (Fourth edition) The Stationery Office, London, 1997.

Acknowledgements
Reproduced by permission of the Building Research Establishment Ltd.

The majority of the work presented in this paper has been sponsored by the UK Department of the Environment, Transport and the Regions (DETR).

15 DEVELOPMENT OF REQUIREMENTS FOR SPECIALIST FOUNDATIONS FOR SPORTS STADIA

P. A. KINGSTON and W. G. K. FLEMING
Kvaerner Cementation Foundations, Rickmansworth, UK

SUMMARY: This paper describes the significant development in piled foundation requirements for sports stadia since the introduction of the Taylor Report in 1990. In particular the paper will focus on the Alfred McAlpine Stadium in Huddersfield, the Reebok Stadium in Bolton, Hampden Park in Glasgow and the Millennium Stadium in Cardiff. Technical advances in continuous flight auger (CFA) piling together with design development have delivered economic and practical foundation solutions.
Keywords: Stadia, Taylor Report, design, foundations, piling, development

INTRODUCTION

Following the Hillsborough disaster on 15 April 1989, the Rt. Hon. Lord Justice Taylor was commissioned to carry out an Inquiry to make "recommendations about the needs of crowd control and safety at sports grounds". As a result, the Taylor Report was published in January 1990. This thorough document made a number of clear recommendations for action necessary by stadia proprietors to minimise the likelihood of any further disaster. The most significant requirement was that standing capacity on terracing must be reduced and eventually eliminated for major English and Scottish football clubs and national stadia by August 1994. For all other designated grounds this requirement would be extended to August 1999.

In order to meet this requirement many football clubs have needed to invest in new stands, additional tiers to existing stands or new multi-purpose stadia. The simpler, cheaper solution to fix seats to existing terraces was discounted by many as it would severely reduce ground capacities and revenue. The decision to build a new stadium has often been taken when a club's original ground is located in a city centre where space for expansion is limited and very costly. Sale of city centre land for building and shopping centre development has helped funding for these projects together with the more recent availability of brand name sponsorship and grants from local authorities.

When designing new stadia the focus has been to provide buildings that will provide a service to the local community and not simply be a venue for fortnightly football matches. In addition, developers have tried to incorporate the philosophy that the supporter is a customer whose needs must be catered for. This principle has been recognised and applied successfully in North America for a number of decades. For financial success any new development must also be multi-functional, flexible and suitable for accommodating a number of different sporting and social events.

Stadia, Arenas and Grandstands, edited by P.D. Thompson, J.J.A. Tolloczko and J.N. Clarke.
Published in 1998 by E & FN Spon, 11 New Fetter Lane, London EC4P 4EE, UK.   ISBN: 0 419 24040 3

STADIA FOUNDATION DESIGN

The history of foundation requirements

In the early 1900s football became increasingly popular in England and Scotland. To accommodate the growing number of spectators, clubs built soil embankments around the sides of the pitch to support terracing. This crude form of support required no foundation design other than a check on slope stability. Such grounds still exist and only recently in 1992 a large clay fill embankment supporting the North Stand at Stamford Bridge, Chelsea was removed and replaced by a modern stand.

Popularity continued to increase and wealthy clubs built roofed grandstands for their spectators which provided shelter from the elements. Designs for stands evolved from cricket pavilions which also housed changing rooms and offices. The foundation requirements for these structures were relatively simple with purely compression loads. Few had major foundations and any piling required relatively limited design.

In the 1950s the next major development came in the form of a fully tensioned cantilever roof. Removal of the regularly spaced columns that supported the roof of early stands dramatically improved the spectators' view and became a popular and economic option. Foundation design remained straightforward but tension loads from overhanging superstructures took on new importance in overall stadia provision. A development on this theme came in the 1970s with the introduction of the goal-post roof where a major beam spanned the full length of one stand, supported at either end by large columns.

With the general decline of attendance in the 1970s and 1980s there was little development of football stadia in the UK. Unfortunately it took the disaster at Hillsborough together with massive television sponsorship to kick-start new growth in construction. Designers have been encouraged to use greater imagination and ensure that the new structures are stylish and appealing. The improved safety standards for crowd control, need for improved versatility and attention to customer care have also led to more complex and demanding requirements from the structure and these needs are reflected in the foundations.

Once again an important development influencing foundations lies in the design of the roof; specifically in the use of corner masts. Foundations are now of vital importance as these supports are required to resist massive horizontal and vertical loading from the structure. Significant structural sway must be avoided and horizontal and vertical strain compatibility must be controlled requiring rigorous analysis.

During the hundred years since the construction of the first football stands the role of ground/structure support of new stadia has changed significantly. Foundations are now of vital importance and in recent years many stadia developments have required major foundation provision.

Recent experience

Fig. 1 shows the breadth of experience gained by Cementation in this field using many of the current piling techniques. The list of over 40 stadia, arenas and grandstands includes:

- Twickenham RFU in London (Rotary bored piling)
- Wembley Stadium in London (Tripod piling)
- Pittodrie in Aberdeen (Driven cast in-situ piling)
- Celtic Park in Glasgow (Precast piling)
- Old Trafford in Manchester (CFA piling)

- Ibrox Park in Glasgow (CFA piling)
- The Oval in London (CFA piling)

During the 1990s over 75% of the stadia piling schemes undertaken by Cementation have used CFA piling techniques. Many of the new sites use land previously considered unsuitable for development due to poor soil conditions or site contamination.

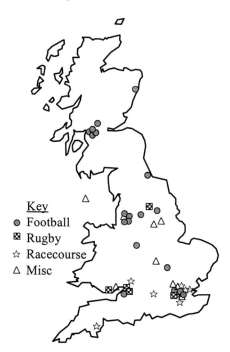

Fig. 1. Plan showing extent of recent stadia projects by Cementation.

Developments in continuous flight anger piling

The process of constructing continuous flight anger (CFA) piles has progressed substantially in recent years. In the UK its general use now represents nearly 40% of the piling market. The technique requires special care during the boring and concreting operations as mis-construction could affect the integrity of the structure. Reliance is placed on instrumentation to ensure a high level of control and give assurance of the quality of installation.

On-board computer based systems display and record the significant actions that affect the successful installation of each pile. This information can then be checked either manually or automatically and statistical results are presented to evaluate rigdriver performance. Theoretical and practical research has been undertaken to determine the best boring technique for various soil conditions. As a result, an electronic control system has been developed to replace many of the manual operations so that pile installation can be controlled automatically by the monitoring computer. Automation of the pile construction has led to improved quality, minimal concrete wastage, shorter installation time and overall cost savings to clients. Piling rigs have also grown in size and power allowing the construction of larger diameter (up to 1200 mm) and deeper piles. Research on auger head configuration has led to improved boring in hard soils and soft rocks. These important improvements in pile construction have given pile designers the confidence that pile installation is reliable and that greater rock penetrations and load capacities can often be achieved.

CASE STUDIES

The Alfred McAlpine Stadium, Huddersfield

Plans for a new stadium in Huddersfield began in 1990. This formed part of a major local redevelopment which has promoted the city on the international stage. Plans were developed not only for the local football and rugby clubs but for numerous other sporting and social facilities including a swimming pool, a golf range, a bowling alley, conference rooms and a creche.

The pile design for the main structure is relatively conventional with the exception of the roof foundations. The roof for each stand is supported by an arch beam running the full length of each side of the pitch. These beams, known as banana trusses, induce large horizontal and vertical loads via four legged finger supports to the cruciform configured piles. Fig. 2 shows the scale of a corner roof foundation in the foreground with the finger supports under construction in the background.

CFA piling commenced in 1993. This technique was chosen as the most practical and economic solution because significant penetrations into the weak Mercia Mudstone were required to ensure that the lateral loads of up to 7000 kN per corner could safely be resisted.

This Stadium has set a high standard for others to follow and is considered to be one of the finest stadiums of its size in Europe. The RIBA described the 'architecture as an economic miracle' when it was awarded the Building of the Year for 1995.

Fig. 2. Longitudinal photograph of project during construction.

Fig. 3. Aerial photograph of the Alfred McAlpine Stadium prior to the recent construction of the Panasonic Stand.

The Reebok Stadium, Bolton

Over the last 12 months Bolton Wanderers FC have seen a number of major changes in their fortunes. These include being floated on the stock market, promotion to the Premier League, major sponsorship deals from television and the private sector and the construction of a new state-of-the-art stadium. The Club's previous home since 1895 was at Burnden Park in the centre of Bolton which held many wonderful memories of football success. General disrepair meant that this aged structure was not suitable for top flight matches.

Situated only 500 m from junction 6 of the M61 motorway so ensuring high visibility, the Reebok Stadium forms part of a 200 acre, £150 million urban regeneration supported by the local council.

Once again the roof support, shown in Fig. 4, required special foundation design considerations. The mast support piles are required to resist group horizontal loads of up to 19,000 kN per corner.

The soil profile across the foot print of the site consisted of competent Coal Measures at shallow depths to the north and very soft peat to depths of over 6 min the south.

Two large bases were required per comer consisting of up to 42 No. 750 mm diameter CFA piles per base. It is important that the foundation movement is controlled and strain compatibility is achieved between the comers. The pile design took account of the soil conditions for each zone and different solutions were adopted for the two northern and southern comers. Site tests were undertaken to confirm the design assumptions.

Fig. 4. The mast legs, Reebok Stadium.

Fig. 5. Aerial photograph of The Reebok Stadium.

Hampden Park, Glasgow

Hampden Park is the home of Scotland's national football team and Queens Park FC who currently play in the Scottish third division. The stadium once boasted to be the largest in the world with a capacity of 150,000. The traditional oval shape has been retained since its opening in 1903 and the design of the structure has remained relatively simple in an effort to commemorate the ground's past.

As a result of the requirements of The Taylor Report the stadium was closed for major construction between 1992 and 1994. During this period 280 No. large diameter rotary bored piles were installed as part of the contract for the roofing of the north and east terraces. At that time heavy duty rotary piling equipment was the best practical solution for overcoming the extremely difficult soil conditions. The moderately strong limestone/ sandstone varied by up to 20 m in level and was overlain by very soft alluvial clays or hard clays with numerous massive limestone boulders.

In 1997 construction recommenced with the rebuilding of the south stand and roofing for the west terracing. The load requirements were very similar to those carried out for the earlier works. An alternative using CFA techniques was proposed as developments in piling technology had been significant during the previous five years. Preliminary trials were carried out which demonstrated the suitability of the proposals.

The foundation element, consisting of 415 No. large diameter CFA piles, was completed during a 14 week period which almost doubled piling productivity as compared with the earlier contract. This was achieved with an additional cost saving to the client of over twenty percent per pile against the previous piling works.

Fig. 6 shows a photograph of a typical modern CFA piling rig with increased torque, auger diameter and depth capability.

Fig. 6. Photograph of a modern CFA piling rig.

The Millennium Stadium, Cardiff

The Cardiff Arms Park, situated on the banks of the River Taff in the heart of the city, is the home of Welsh rugby. Many visiting fans have commented on the wonderful atmosphere generated at this sporting venue during international fixtures. The original structure, although steeped in history, was becoming dated and in need of repair and the Welsh Rugby Union in conjunction with Cardiff City Council devised plans for a new world beating stadium.

The new 72,500 capacity stadium will be located on the same but extended site of the original Cardiff Arms Park. This decision, although welcomed by rugby fans, generated a number of major engineering and logistical challenges. This 'third generation' stadium requiresthe foundations to satisfy an onerous performance specification for the very heavy loads. To ensure that low settlements are achieved large pile penetrations into the zone II Mercia Mudstone are necessary. The tender conforming scheme for the foundations consisted of approximately 1000 No. large diameter rotary bored piles. Once again an alternative scheme, using the most powerful CFA equipment available, provided cost savings to the client of approximately 30%. Two preliminary test piles demonstrated that the proposals were practical and achievable and that the onerous performance specification could be satisfied. Rig instrumentation has proven invaluable in assisting in the identification of the variable level of the founding strata. This information has been presented as part of the pile installation record. The main piling commenced in June and was over 75% complete by the end of 1997, approximately two weeks ahead of the contract programme.

Fig. 7 Aerial photograph of the Millennium Stadium during construction.

A retractable roof has been incorporated into the stadium design to provide greater versatility. Once again the entire roof will be supported via masts on piled foundations located at the comers of the pitch. The horizontal loads from the roof have increased with the size of the stadium and have been calculated as up to 26,000 kN per corner.

A number of options were considered by the project team including the use of a prestressed ground anchor system, 1500 mm diameter rotary bored piles and raking rotary piles, all in various configurations. Each option was analysed and cost budgets provided. The optimum economic solution consisted of 900 mm diameter rotary bored pile groups with three rows of raking piles at 1 in 6 to the vertical and four rows of vertical piles. The outline of the completed north-west mast base is shown in the top right comer of the aerial site photograph in Figure 7.

Careful planning and teamwork by each member of the construction team has been essential in overcoming major engineering challenges which have also included piling adjacent to sensitive structures, working in restricted conditions, the careful sequencing of other construction activities during piling and the need for a sympathetic approach to construction for this City Centre site.

CONCLUSION

There has been a significant increase in the development of UK stadium design and construction following the recommendations of the Taylor Report. The need for multipurpose all seater stadia and the demand for attractive structures has allowed Architects and Consulting Engineers to devise new concepts in design. Foundations requirements have increased in complexity and are now a greater proportion of the construction costs.

Suitable sites for these large structures often have either poor soil conditions, contamination due to previous site history or existing buildings requiring demolition. Economy throughout the construction and building life is essential to ensure the project is commercially viable and foundations have taken on new importance. The significant developments in foundation construction, in particular CFA piling, and design have played a large part in driving out cost saving solutions.

Future economies will be made with improvements in design technology as current methods have been pushed to their practical limits. A new era in non-linear pile group analysis is currently under development but this must be matched with improved methods of site investigation. Early identification of foundation requirements is essential in order to maximise these benefits.

REFERENCES

1. Rt. Hon. Lord Justice Taylor, The Hillsborough Stadium Disaster, Home Office, HMSO, 1990.
2. England, M. New techniques for reliable pile installation and pile behaviour design and analysis, TRB Conference Washington, January 1994, Paper No. 94542.
3. Dickson, M. G. T. Towards safer stadia, The Structural Engineer, Vol. 69, No. 19 October 1991.

PART FOUR

SERVICES AND ENVIRONMENTAL DESIGN

16 ENVIRONMENTAL LOADS AND MICROCLIMATE IMPACTS ON SPORTS FACILITIES

M. J. SOLIGO, J. B. LANKIN and P. A. IRWIN
Rowan Williams Davies and Irwin Inc., Guelph, Ontario, Canada

SUMMARY: Wind loads, snow loads and the microclimate in sports facilities affect their cost and success. This paper describes current methods for determining environmental loads and the microclimate for the purposes of structural design, cladding design, designing the spectator comfort, and promoting turf growth. The methods employ a range of techniques including wind tunnel tests, Computational Fluid Dynamics (CFD) and other computer simulation techniques. Examples of specific applications are given, several being retractable roof stadiums. For these structures, the governing load for the roof drive system is usually wind. The application of Computational Fluid Dynamics methods in the design of the air conditioning system of the Bank One Ball Park in Phoenix is described. CFD methods were also used to assess rain infiltration in the new baseball stadium in Seattle. In the case of a new baseball stadium in Milwaukee, snow load studies are discussed.
Keywords: stadiums, wind, snow, loading, rain, testing, simulation, comfort.

INTRODUCTION

Modern stadiums and arenas are a large financial commitment, making it important to have an economical design. At the same time, the public and owners have high expectations of the finished product. Besides looking good and functioning well from the point of view of sight-lines and acoustics, there must be no question as to its structural integrity and safety under various environmental loading conditions such as wind, snow and earthquake. Since structure is a large part of the cost, accurate knowledge of these environmental loads is important. The microclimate within the facility is also important in so far as it affects the comfort of spectators and the sports activities themselves. Therefore, good knowledge of the internal microclimate is needed including: internal wind flows; solar radiation; shadow patterns; temperature; humidity; and rain infiltration.

This paper reviews current methods that are used to obtain the environmental loads due to wind and snow, and to obtain detailed information on the microclimate. Examples of specific stadium projects are described. Wind tunnel tests on scale models form a central part of these methods. Noteworthy is the variety of shapes that present day designers are coming up with. Wind and snow loads are highly sensitive to shape.

Stadia, Arenas and Grandstands, edited by P.D. Thompson, J.J.A. Tolloczko and J.N. Clarke.
Published in 1998 by E & FN Spon, 11 New Fetter Lane, London EC4P 4EE, UK.   ISBN: 0 419 24040 3

In addition to the wind tunnel, computer simulation is playing an increasing role in these types of study. For example, hybrid wind tunnel/computer simulation techniques have been developed for predicting snow loads including cumulative snow drift and melting effects, and also for assessing spectator and pedestrian comfort. In both these applications, the wind tunnel's unique capability of accurately simulating the wind flows around complex shapes is combined with the power of computer simulation methods in tracking multiple variables such as snowfall, temperature, solar radiation, shadows, etc., which are not readily simulated directly in the wind tunnel.

MICROCLIMATE ISSUES IN DESIGN

Typical issues that affect the design of a sports facility and the methods used to address them are discussed below.

Wind loading on the main structural systems
Wind loads on the main roof system, scoreboards, and lighting systems. These loads are typically determined through boundary layer wind tunnel tests on a rigid model [1], although preliminary estimates are often made based to approximate calculation to help early design iterations.

Wind loading on cladding
The wind loads on the cladding of the roof and walls are typically established by boundary layer wind tunnel tests [2].

Wind-induced vibrations
Large span roofs can become subject to vibrations or oscillations due to their flexibility. Aerodynamic stability needs to be assured. In cases where stability may be an issue, wind tunnel tests on an aeroelastic model (a flexible model) are typically undertaken [1].

Wind loads on roof drive and braking systems
For operable roofs, the power requirements for moving and braking the roof are usually governed by the wind drag loads. Uplift may also be important in some cases. These loads are best evaluated by wind tunnel studies on either a rigid or a flexible model representing various stages of the opening process. Limiting loads for operation of the drive system are usually selected to have a recurrence interval of between about one and five years.

Snow loads
Large span roof designs in areas subject to cold winters can be sensitive to snow loads, including unbalanced and concentrated loads created by drifting. Snow loads are established through a combination of scale model tests and computer simulations [3, 4].

Spectator/pedestrian comfort
Wind can have adverse effects on spectators, particularly in colder climates. In hot climates solar radiation, shadows, clothing levels, activity, temperature, and humidity can be equally important in affecting comfort and more, rather than less, wind may be a desirable feature. A comprehensive assessment of all these factors is possible in terms of frequency of

occurrence of uncomfortable conditions by a combination of wind tunnel studies and computer simulations [5].

Natural turf
Natural turf needs the right ventilation and other microclimate conditions to survive. The frequency distribution of the wind speed over the turf as well as that of other parameters such as solar radiation, cloud cover, dew, precipitation and humidity can be predicted by a combination of wind tunnel and computer studies

Wind effects on the play of the game
Games such as baseball, North American football, soccer, rugby, and some track and field events can be adversely affected by strong wind currents within the stadium. Sometimes, for example, the effect of wind on the trajectories of baseballs may be of interest. An assessment of these effects is possible through wind tunnel tests and computer simulation studies.

Rain infiltration
Knowledge of the extent of rain infiltration rates through openings at the edges of roofs or under cantilever roofs, along with frequency of occurrence, is sometimes needed. The rain drop trajectories are best predicted with the aid of Computational Fluid Dynamics (CFD) studies [6], supplemented by wind tunnel tests to provide boundary conditions such as wind velocities or pressures at the perimeter of the computational domain.

Mechanical heating and air conditioning
Since stadiums and arenas have large internal volumes, the variations of air temperatures and velocities within the volume can be very significant. What is important is to ensure the conditions are comfortable in all the areas occupied by spectators and players, i.e. in the seats and on the field. Frequently the detailed design of the mechanical systems can be improved and optimized with the aid of CFD studies [6,7], leading to better value for dollars spent on the mechanical systems, and perhaps, reduced capital and operating costs. The transient performance of the mechanical systems may need to be evaluated (i.e. in bringing the stadium from ambient to comfortable conditions in a short period) as well as the steady state conditions.

Noise and acoustics
Preventing excessive noise inside the facility may be important in some cases, especially if located near an airport or major transportation artery. Also, good acoustics may be required inside for concerts or similar events. Therefore, input from noise and acoustics experts, early in design, is important.

Air quality
Contamination of air by exhausts from traffic or stationary sources may give rise to problems in some areas (e.g. bus drop off zones, cooling towers, boiler exhausts). Typically an initial screening of various sources by approximate computational methods is sufficient to identify potential problem areas. If solutions are proving difficult to develop, then more detailed and accurate information from wind tunnel tests using tracer gas and flow visualization methods are helpful.

CASE STUDIES

Case Study 1: Paul Brown Stadium, Hamilton County, Cincinnati, Ohio

This facility is currently under design, the primary purpose being for North American Professional Football. A photograph of the wind tunnel model of the proposed facility is shown in Figure 1. The architects for the stadium are NBBJ of Los Angeles, California, the structural engineers are Ove Arup of Los Angeles, California, and the turf consultant is Turf Diagnostics of Kansas City, Missouri.

Fig. 1. Wind tunnel model of Paul Brown Stadium.

Of concern for this structure was wind and snow loading on the large 26m wide canopies/sun screens. Two types of canopy designs were considered: light weight metal; or a fabric surface. Wind and snow loads were therefore evaluated through model tests and computer simulation. Use of louvres at the canopy was explored as a means of reducing wind loads. In addition to wind loading on the main structural system, local cladding loads were evaluated, see Figure 2. As well as the usual spectator and comfort studies, a fairly detailed evaluation of the microclimate was undertaken from the point of view of the turf.

The turf consultant was interested in shadow patterns on the playing and practice fields. To provide this information, sun and shadow patterns were prepared for every hour of the day on March 21, June 21, July 31, September 21, October 31 and December 21. This included the Summer and Winter solstice representing the extreme patterns, the Autumn and Spring Equinox representing average or middle patterns, and two additional dates requested by the consultant. A sample rendering for March 21 at 8:00 am is shown in Figure 3.

While this information provided sun and shadow patterns, more detailed information was required on actual predicted hours of solar radiation reaching the playing field. This

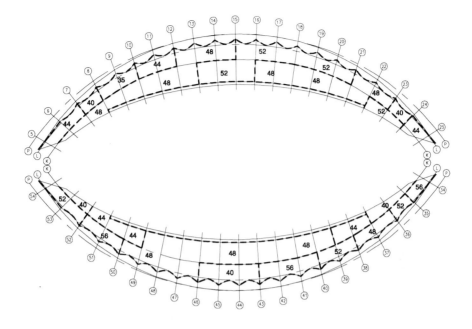

Fig. 2. Cladding loads.

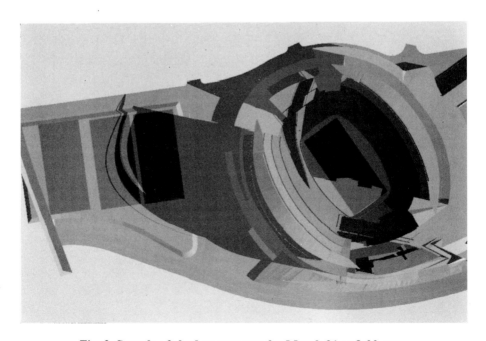

Fig. 3. Sample of shadow patterns for March 21 at 8:00 am.

was determined by conducting a meteorological analysis of Global Hourly Radiation (GHR) and Opaque Sky Cover (OSC) data recorded at Covington Greater Cincinnati Airport from 1980 to 1990.

GHR is the total amount of direct and diffused solar radiation in Wh/m^2 received on a horizontal surface. This data was combined to provide average cloud cover and solar radiation given the actual hourly weather conditions. Radiation amounts in Wh/m2 were converted to Equivalent Hours of Sunshine (EHS) per day. This is a unit of measure which defines the maximum amount of solar exposure a given surface would receive relative to the maximum amount of solar exposure an unobstructed surface would receive at solar noon or June 21, assuming a clear day. For this study, the EHS was 963 Wh/m^2 per hour as calculated using ASHRAE procedures [8].

A three-dimensional computer model was constructed of the stadium and surroundings. Forty-five specified locations were set on the playing field and hourly measurements on the 15th day of each month assessed as to sun and shadow exposures. This data was then combined with the meteorological assessment to determine the EHS at each location. The data was then presented in Tabular and Graphical Contour forms. An example of a contour plot is provided as Figure 4.

The results of the Solar Radiation Study and wind tunnel results on wind speeds over the field provided the turf consultant with sufficient data to set turf specifications including turf type, growing potential, and projected replacement timing.

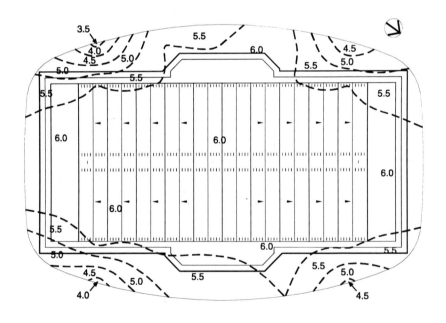

Fig. 4. Contour plot of equivalent hours of sunshine per day.

Case Study 2: Bank One Ballpark, Phoenix, Arizona

This facility will open for baseball in April of 1998. A photograph of the wind tunnel model is given in Fig. 5, showing the stadium with the roof partially open. The architects are Ellerbe Becket of Kansas City, the structural designers are Martin & Martin of Denver, Colorado, and the Mechanical Designers are M-E Engineers of Denver, Colorado.

Phoenix's environment is dry and hot throughout the baseball season with summertime temperatures often reaching 110°F (43.3°C). To provide a comfortable experience that will attract baseball fans during the summertime, a retractable roof stadium was decided upon. In this way, the roof could be closed and the stadium air-conditioned on hotter days, but left open for cooler conditions. A retractable roof also allowed for the use of natural turf. Natural turf is preferred for both nostalgic and aesthetic reasons, and because players are less likely to be injured. Additionally, six large doors making up approximately 50% of the north facade of the stadium were incorporated into the design to provide natural ventilation at times.

The microclimate and loading studies included: wind loads for structural and cladding design (with roof in various positions); an assessment of wind induced roof vibration; wind loads on the roof drive system; CFD studies of the air conditioning flows including the thermal lag effect of the stadium structure; spectator comfort; and wind effects on baseball trajectories.

For the wind loading studies, a model of the stadium and surroundings was constructed to a scale of 1:400. In order to assess the impact of the various roof positions and north wall door openings on the structural and cladding wind loads, the stadium model, was tested in RWDI's 1.9 × 2.4 m (cross-section) wind tunnel for the following six test configurations:

1. Main roof closed, North doors closed
2. Main roof closed, North doors open
3. Main roof partly open, North doors closed
4. Main roof partly open, North doors open
5. Main roof open, North doors closed
6. Main roof open, North doors open

The model was instrumented with a total of 743 pressure taps located on both the exterior and interior surfaces of the stadium. In this manner, direct measurements of the net wind pressures (mean and gust) acting across roof and wall surfaces, were taken, for all but Configuration 1. For Configuration 1 the measurements on the exterior pressures were taken and later an internal pressure estimate was added to these measurements. To determine the net design wind loads, the wind tunnel measurements were combined with a meteorological analysis of the Phoenix winds using the up crossing technique[2] for each of the six configurations. The results for each configuration were then compared and the worst case results were provided. In the case of the structural wind loads on the roof, the impact of the roof dynamic response in the lower modes of vibration was included in the loads.

To examine the possibility of unstable roof oscillations occurring and to obtain more accurate evaluations of the horizontal drag loads on the roof panels, a 1:150 scale aeroelastic model was built. This flexible model was designed to react to input forces and move in a manner similar to the full scale roof. The model was fixed to a rigid stadium base and tested in RWDI's 3.0 × 4.9 m (cross-section) wind tunnel to determine the following information:

Fig. 5. Bank One Ballpark with roof open.

1. Aerodynamic stability of the roof panels was assessed for various wind angles, wind speeds and test configurations to ensure that the aerodynamic damping of the roof was positive. This was also used to check the assumption that the set of wind loads measured on the rigid model (Structural Wind Load Study) were not adversely effected by roof motions, or, if they were, to provide the necessary adjustment for them.
2. Using laser deflection measurement devices, the roof deflections were measured to help determine the gap sizing between roof panels.
3. The roof panels were also strain gauged in order to measure the drag wind loads affecting the operation of the roof mechanical drive system. The drag loads were provided as mean and gust component loading so that loading over various averaging times between 0.2 s and 1 h could be assessed in the design of the roof drive systems.

The Stadium's mechanical engineers were concerned about how to accurately and economically assess the cooling load in the stadium during a four hour cool down period prior to game time. Turf growth required that the roof stay open until 3 pm for a 7 pm game to maintain the health of the grass. This means that the majority of the materials (i.e. concrete seating decks, plastics, vinyls, etc.) comprising the inside of the stadium, will be in direct sunlight throughout the day, absorbing heat and raising the surface temperature of the materials (i.e. concrete to 140°F or 60°C) by mid-afternoon. An air conditioning system was required to cool the stadium down between the roof closing at 3 pm and the 7 pm game start time. The mechanical engineers knew they could provide enough cooling, but did not know how much was required to account for the heat release from the materials, or whether the materials would cool sufficiently to avoid the condition where spectators would have hot feet and backsides from the material heat release, and cool heads from the airconditioning. To meet this need, RWDI developed an approach for predicting this cooling load. This method also predicted concrete temperatures and air temperatures in the stadium throughout the cool down period and at game time. This was achieved by using techniques that included: wind tunnel testing, one dimensional transient thermal modelling, meteorological analysis, field studies, sun/shadow three dimensional modelling, and CFD modelling.

To provide boundary conditions for the CFD model, the 1:400 scale model of the stadium was tested in the wind tunnel. Hot wire anemometry techniques were used to measure the velocity of air entering and leaving the proposed openings in the ends of the stadium, and flow visualization was used to confirm the airflow patterns.

To assess the heat storage and release in the concrete used in the seating bowl of the ballpark a one-dimensional transient thermal numerical model was developed. The model was applied at several key locations representing areas of the deck that would experience different external conditions. Meteorological data was used to evaluate typical conditions in Phoenix in the middle of hot summer months. Based on the hourly temperature data for a number of typical summer days, a set of 24 hourly temperature representing the ASHRAE 99% design point was produced and incorporated into the one-dimensional model. To provide confidence in the selection of the appropriate heat transfer properties for the concrete, field measurements were undertaken for two days in the seating section of the Sun Devil Stadium at Arizona State University. Scenarios of sun and shade for several locations within the seating bowl were formulated for daylight hours using a three dimensional CAD model of the stadium with the roof open. By viewing the stadium from the sun's vantage point for particular dates and times, areas in the sunlight and shadow were determined. The model also incorporated the effect of the seats providing localized shading of the concrete immediately under them. This solar flux information was incorporated into the one-dimensional model.

Variations in concrete and air temperatures over the course of the day at representative locations were predicted. The results showed that, because of time lags in the 100 mm to 130 mm thick concrete, the thermal history over the previous six hours was significant. The hollow lightweight seats provided significant shading for the concrete which reduced the temperatures and improved comfort.

A CFD model was developed to predict spatial airflow patterns and temperature distributions in the ballpark which incorporated the predicted heat release of the concrete, as well as, the estimated heat load of spectators and lighting. It allowed the effectiveness of the design of the conditioned air delivery system, in terms of providing uniform comfortable conditions at game time to be assessed. The transient heat flux models were modified by the results of the CFD modelling, which in turn provided new inputs for the CFD model. This iterative procedure was carried out several times.

Outputs from the modelling provided time histories of the cooling load throughout the stadium, as well as, the surface and core temperatures of the concrete. The CFD modelling also provided air temperatures and speed distribution. The results showed that the negative buoyancy of the chilled air required re-directing of the supply jets upward from their original direction. In order to provide the best balance between delivery of cool air to the farthest seating and avoid cold drafts near the jets, they were aimed horizontally instead of parallel to the seating. Areas were identified near the openings to concourses where ventilation could be reduced at times to cut operating costs while still maintaining comfort.

Other recommendations as a result of this study included: closing in proposed openings at the ends of the stadium so wind-driven air currents would not adversely affect the distribution of conditioned air; continuous control of the roof opening size and position, so the amount of sunlight received by the grass could be optimized and the concrete heat gain could be reduced; and the use of local shading devices to solve overheating problems in smaller areas of the stadium.

Case Study 3: New Pacific Northwest Ballpark, Seattle, Washington

This baseball facility is currently under design and construction, and is scheduled for opening in April of 1999. A photograph of the aeroelastic wind tunnel model of the proposed facility with the roof closed is shown in Figure 6. The architects are NBBJ of Seattle and the structural engineers are Skilling Ward Magnusson Barkshire Inc. also of Seattle.

As with the Phoenix stadium in the preceding case study, wind loading on the structure and cladding, wind induced vibration, wind loads on the drive system, spectator comfort and the effects of wind on the baseball were studied.

The Seattle environment is mild and during parts of the baseball season is also wet. As a result, a retractable roof was designed for this facility. However, the stadium was designed to reflect an open air concept so the roof acts more as an 'umbrella' with the sides of the facility open. One particular concern was the extent of wind driven rain infiltration into the facility. Also being in a climate where snowfall is common in winter, snow loads were an additional concern. Since snow loading is discussed for Case Study 5, attention here will be focussed on the rain infiltration study.

The rain infiltration study was conducted using an analysis of wind and rain data from meteorological stations in the Seattle area, and employing Computational Fluid Dynamics (CFD) techniques to determine the trajectory of the raindrops.

An analysis of three meteorological stations in the Seattle area was performed first to assess the directionality of combined wind and rain events for the months of the baseball

Fig. 6. New Pacific Northwest Ballpark with roof closed.

season, April through October. This analysis indicated that winds from the southeast through west were the primary directions for joint occurrence of wind and rain.

For the important wind directions, wind tunnel measurements of boundary conditions were made. A three-dimensional computer model of the stadium was then constructed for use with the CFD analysis. Computational Fluid Dynamic techniques were then used to examine the extent of the rain impact inside the stadium for three rain droplet diameters (1, 2 and 3 mm) which corresponded approximately to light, medium, and heavy rainfalls, see Figure 7. A series of wind speeds ranging from 4 m/s to 16 m/s were also examined for each raindrop size. Where rain infiltration occurred causing wetting of the stadium seating areas, various forms of rainscreen/wind deflectors were developed. These mitigative screens were attached to the physical wind tunnel model and the boundary conditions of flow around the stadium were revised. Using updated boundary conditions, the CFD analysis was rerun with the computer model screens also in place. In this manner, wind deflectors and rainscreens were developed for the stadium roof design to minimize the rain infiltration onto the spectator seating areas.

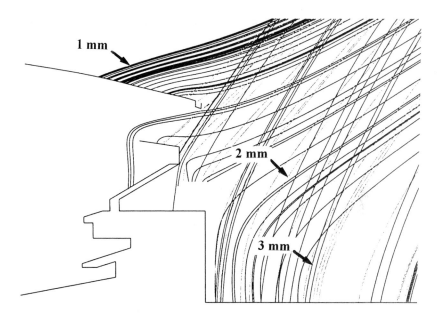

Fig. 7. Example computation of rain drop trajectories.

Case Study 4: Hong Kong Stadium, Hong Kong

This facility, Figure 8, serves as the venue for the annual World 7's Rugby Championship among other sporting uses. The architect was Hellmuth, Obata and Kassabaum and structural engineers, Ove Arup & Partners. Two large sawtooth shaped sun canopies cover the main seating areas and the stadium is located in a valley which channels winds.

The wind tunnel studies, Figure 9, focussed primarily on wind loading of the canopies and the influence of surrounding hilly topography.

In order to determine the effects of topography on the wind speed and turbulence characteristics at the project site, wind tunnel tests were conducted on a 1:3000 scale

Fig. 8. Hong Kong Stadium

topographic model of the area, which included the surrounding features within 7.5 km of the stadium site. The model was tested in RWDI's 4.9 m wide wind tunnel. The mean wind speed and local turbulence intensity profiles above the stadium site were measured using a hot film sensor for a variety of wind directions. The results from the 1:3000 scale model were used to tailor the flows approaching the 1:300 scale model in the wind load tests.

The wind tunnel results identified unusual unbalanced loads on the sun canopies caused by local terrain. For south winds, a major portion of canopy was subject to negative (upward) loads while the remaining portion was simultaneously subjected to positive (downward) loading. The tooth canopy design also was found to result in significant cumulative drag loads for wind parallel to the canopy's long axis. Wind loads for the design of individual valley cables and trusses on the roof panels, as well as for larger tributary areas of fabric, were determined.

Case Study 5: Miller Park, Milwaukee, Wisconsin
This facility is currently under design and construction. The architects are NBBJ of Los Angeles and HKS Inc. of Dallas, and the structural engineers are Ove Arup, Los Angeles. The proposed stadium has a retractable fan shaped roof. Figure 10 shows the aeroelastic model under test. The issues studied included: wind loads on cladding and structure, wind vibrations, wind loads on the drive systems, snow loading, spectator comfort, and effects of wind on the baseball. Since Milwaukee experiences substantial snow loads, this aspect was important and will be described here. The retractable roof enables natural turf to be used. To promote proper growth of the turf in the spring the stadium roof may be either left open during the winter or opened in the spring as long as the snow on the roof does not restrict the movement of the roof panels. A 50 year return period snow loading was determined for the roof in both the open and closed positions.

Fig. 9. Hong Kong Stadium model in wind tunnel.

The snow loading was determined for each roof position using a two stage approach which is described in more detail in References 3 and 4. The first stage involved constructing a 1:400 scale model of the roof and testing it in an open channel water flume. To simulate snowfalls, sand particles were introduced into the waterflow upstream of the submerged model. This method of simulating snowdrifting on the scale model is used as a guide to the presence of unusual drift formations. The second stage was a more detailed snow load study referred to as the Finite Area Element Method (FAE). This method integrated the results of the water flume testing with detailed wind tunnel testing, and a computer simulation including meteorological data. The detailed approach using the FAE method was developed initially for work on the Toronto SkyDome. Since that time it has been used to establish design snow loads on many other structures and to conduct snow load research on large buildings for the Canadian Building Code.

Most of the data required as input for the computer simulation was obtained by analysing the local meteorological data. This data included detailed wind, snowfall, rainfall, air temperature, relative humidity, and cloud cover records. To obtain the local wind velocity patterns over the roof, scale model wind tunnel tests were performed. The 1:400 scale model of the stadium used for the flume testing was instrumented with 166 wind speed sensors for the closed position and 92 sensors for the open position and tested in RWDI's boundary layer wind tunnel. These tests provided measurements of wind speed and direction at various points on the roof. Since winds are accelerated around corners and over ledges, the effect of wind on roof level snow accumulations is not uniform.

Snowdrifting, due to wind, increased snow loads on some segments of the roof and decreased snow loads on other segments. The FAE computer program combined the meteorological data with the wind tunnel results to predict the distribution of snow loads on the roof. The effects of snowfall, snowdrifting, retention of water from rainfall, and reductions due to melting are all accounted for. The FAE simulation used 29 years of detailed meteorological data from Milwaukee's Mitchell Field airport, simulating each winter of those years in hourly intervals. Load cases important for the structural design were developed, including uniform and unbalanced loading on the whole roof, individual panels, and panel halves, as well as step loading adjacent to the vertical edges of each panel.

Figure 11 shows an example of one of the resulting design snow load distributions for the roof in the closed position. Note that, in addition to 50 year loads, the more typical snow accumulations at the end of the winter were also evaluated to assist in addressing the question of when the roof could normally first be operated without interference from snow accumulations.

Fig. 10. Miller Park aeroelastic model in wind tunnel.

CONCLUSION

The combination of scale model studies and computer simulation methods allows a more scientific approach to be taken to determine the design loads and microclimate of modern sporting facilities than was possible in the past. Since wind and snow loads often drive the structural design of the large span roofs, these facilities, of more accurate knowledge of these loads provides significant benefits. Other issues such as spectator comfort, the effect of wind on the events in the stadium, and the microclimate for the turf are also areas of concern for the designers. Again, there are techniques available to help the designer with arrive at effective solutions to problems in these areas.

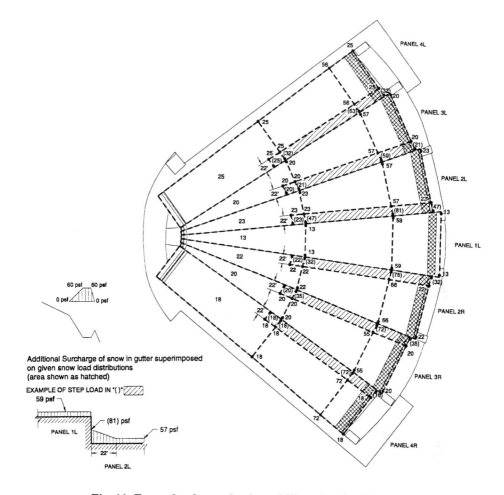

Fig. 11. Example of snow loads on Milwaukee Stadium.

REFERENCES

1. Grieve, J. B., Stone, G. K., Irwin, P. A. Prediction of structural wind loads on the large span roofs, *Innovative Large Span Structures*, Vol.1, pp.404-414, IASS-CSCE International Congress 1992 .
2. Irwin, P. A. Pressure model techniques for cladding loads, *Journal of Wind Engineering and Industrial Aerodynamics*, 29, pp.69-78, 1987.
3. Irwin, P. A., Gamble, S. L. Model and computer studies of snow loading, *Innovative Large Span Structures*, IASS - CSCE International Congress, Toronto, Ontario, Canada 1992.
4. Irwin, P. A., Gamble, S. L., Taylor, D. A. Effects of roof size, heat transfer, and climate on snow loads: studies for the 1995 NBC, *Canadian Journal of Civil Engineering*, 22, pp.770-784, 1995.

5. Soligo, M. J. Irwin, P. A., Williams, C. J., Schuyler, G. D. Pedestrian comfort: a discussion of the components to conduct a comprehensive assessment, *Eighth U.S. National Conference*, Baltimore, Maryland 1997.

6   Sinclair, R. J. Irwin, P. A., Matson, K. M., Vanderheyden, M., Kriksic, F. Applications of CFD flow modelling in building design, building an international community of structural engineers, *Proceedings of Structures Congress XVI*, Structural Division/ASCE, Chicago, Illinois, April 1996.

7. Gamble, S. L., Sinclair, R. J., Matsui, K. M., Barrett, M. R. D. Ventilation requirements for New Phoenix Ballpark, *ASME Fluid Engineering Division Summer Meeting*, San Diego, California, 1996.

8. 1997 ASHRAE Handbook Fundamentals (SI Edition), *Fenestration*, Chapter 29.17, American Society of Heating, Refrigerating and Air-Conditioning Engineers, Inc., Atlanta, GA.

17 THE USE OF WINDBREAK FENCING TO CONTROL ATMOSPHERIC WIND CONDITIONS IN SPORTS STADIA

M. PICK
Linear Composites Ltd, Keighley, UK

SUMMARY: This paper considers the effectiveness of porous windbreak fence systems in controlling wind speed. It then looks briefly at options for constructing windbreak fence walls before focusing on the use of Paraweb Windbreak Fence, a synthetic composite material which offers unique design and build opportunities. Two case history examples are reviewed, which represent opposite ends of the spectrum for windbreak wall complexity. In both cases, real benefits have been gained for competitors, spectators and the managers of sporting complexes.
Keywords: composite materials, costs, foundations, porosity, prevailing winds, tennis stadia, windbreak fencing, wind speeds, wind tunnel modelling.

INTRODUCTION

The high standards demanded by competitors in today's sporting events are matched only be the high aspirations of the spectators. Increasingly it is the responsibility of stadium designers and managers to ensure that conditions within the competition arena are ideal. At the planning stage the effects of atmospheric wind on events are often overlooked to the detriment of spectators, competitors and ultimately the commercial success of the venue.

An example from 1995 serves as a case in point. Jonathan Edwards competed a world record distance triple jump at Lille of 18.43 m only to have it disallowed because the wind speed exceeded the maximum for record ratification by 0.4 m/s. Happily he went on to set a new record of 18.29 m. Lille, however, missed the opportunity of being the world record breaking venue.

In less critical, but often more severe conditions, high wind speeds, gusts and wind eddies can disrupt and spoil sporting events of all types. The use of well-designed but relatively low cost windbreak fences can help to resolve these problems and provide ideal conditions for competitors together with increased comfort and enjoyment for spectators.

POROUS WINDBREAK SYSTEMS

The basic requirement of a windbreak fence is that it should interfere with the free stream wind flow, producing an area of relative calm in its lee but without generating turbulence.

Stadia, Arenas and Grandstands, edited by P.D. Thompson, J.J.A. Tolloczko and J.N. Clarke.
Published in 1998 by E & FN Spon, 11 New Fetter Lane, London EC4P 4EE, UK.   ISBN: 0 419 24040 3

To be acceptable, it must also be aesthetically pleasing and cost-effective.

The following examples illustrate how porous windbreaks work:

- Placing a solid wall (0% porosity) in a 'free stream' wind flow causes the flow to rise over the windbreak, creating an area of low pressure, a partial vacuum, to the leeward side. The partial vacuum pulls the free stream flow downwards and the windbreak effect persists for only a short distance. The situation can be worse as friction in the boundary region between the free stream flow and vacuum area can generate short-term, often high-speed, turbulence which can be carried to ground level. Wind velocity in these turbulent regions can be higher than those in the free stream. Solid walls do not make effective windbreaks.
- By introducing porosity into the windbreak wall, a diffuse flow of wind passes through into the leeward region. This diffuse flow has significantly less energy than the free stream but is reasonably close to equilibrium with the free stream, so the downward suction effect is much less. The result is wind shelter which persists for a greater distance which, except in extreme conditions, avoids the development of high-speed turbulence.

The optimum porosity for a windbreak wall has been established by several authors [1–3] to be in the region of 40–50% porosity. The effect is summarised in Table 1.

Table 1. Wind speed behind barriers of vertical height h, expressed as a percentage of the upstream wind speed flow, shown in multiples of the fence height (h) downwind of the barrier. After Prior and Keeble [1].

Distance from windbreak fence	0 h	2 h	5 h	10 h	15 h	20 h	25 h	30 h	40 h
Type of barrier:									
Open (porosity 70%)	90	80	70	75	85	90	95	100	100
Medium (porosity 50%)	40	25	20	25	50	60	75	90	100
Dense (porosity 0%)	0	20	40	65	80	85	95	100	100

WINDBREAK TYPES

There are many types of windbreak wall structures available; they range from simple open brickwork, through punched steel sheeting, to synthetic composite materials.

The use of traditional materials, such as brickwork or steel sheeting, has several major drawbacks:

- The materials are heavy and therefore require massive foundations and supporting structure. Erection may also require major plant.
- The materials are intrinsically rigid and fail in bending by fracture or buckling and as a result support pillars or posts have to be placed relatively close together.
- Rigid windbreaks with sharp edges can generate turbulence.
- Small hole sizes, found with some punched sheet windbreaks, can suffer from blinding.

Synthetic windbreaks in the form of drawn polymeric mesh offer some advantages over traditional materials. They are light in weight and relatively easy to install and due to their lower stiffness can move and bow in the wind, avoiding the generation of turbulence from sharp, rigid edges. However, their tensile strength and stiffness are usually relatively low and, as a result, close post spacing is required. Hole or gap size also tends to be small, and in some applications, this can cause problems with blinding, for example, from waste materials or snow.

Synthetic composite windbreaks offer the advantages of synthetic windbreaks with the addition of increased strength and axial (along the plane of the fence) stiffness. This allows higher tensioning forces to be used during assembly, producing a more stable windbreak fence structure. The higher strength windbreak cladding material also allows wider post spacings to be used. The higher axial tension and inherently higher stiffness reduces the amount of deflection or bowing experienced under wind loading. Excessive bowing is not desirable as it may cause mechanical interference and can lead to abrasion or fatigue damage of the fence cladding material. The dynamic response of synthetic composite wind-break fences, bowing rather than buckling, allows load to be transmitted axially along the windbreak fence. This load is balanced on either side of the intermediate posts and can be controlled at the ends of the fence by suitable strengthening or guying of the end posts or by the use of rigid bracing between posts.

Table 2 provides a rough cost comparison for the construction of a 10 m-high permanent windbreak fence manufactured from a range of materials. Posts and foundations represent 50 to 80% of the fence system cost; rigid fence materials (metal plates and masonry) are therefore relatively expensive as short post distances are required. Similarly, open net windbreaks require close post spacing due to the relatively low tensile strength of the cladding material.

Table 2. Relative cost per square metre of windbreak fences fabricated from a range of materials, with Paraweb Windbreak Fence (Kevlar® 29) as the reference.

Windbreak fence type	Post or column spacing (m)	Relative cost
Masonry	10	163
Punched galvanised steel	7	140
Punched aluminium sheet	7	146
Open net synthetic windbreak	3	130
Paraweb Windbreak Fence (polyester fibre, 1700 kg web strength)	16	104
Paraweb Windbreak Fence (Kevlar® 29 fibre, 2000 kg web strength)	20	100

Steel posts have been assumed for all fences other than the masonry fence for which reinforced concrete was assumed. The fence structures have a design wind speed of 45 m/s. Information supplied by Biostructures srl [7].

PARAWEB WINDBREAK FENCE

Paraweb Windbreak Fence is a synthetic composite windbreak fence which is fabricated from extruded polymeric webbing reinforced with parallel lanes of industrial grade polyester or high modulus, high strength Kevlar® 29 fibres. The individual 50 mm-wide webs are formed into windbreak fence panels which have a porosity of 50%. The resulting Paraweb Windbreak Fence structure is lightweight, durable and easy to handle. Individual web breaking loads can be as high as 3000 kg.

The load-carrying and deflection behaviour of Paraweb Windbreak Fence panels are defined by the response of the reinforcing fibres. The use of polyester reinforcing fibres offers a low cost and effective solution for smaller fences in areas where reasonable amounts of fence bowing can be tolerated. Kevlar® 29 reinforcing fibres are used in large fence systems where full benefit can be taken of their higher strength and axial stiffness. The behaviour of both fence types, including their elastic and viscoelastic behaviour, are well understood [4].

Although the strength and stiffness of Paraweb Windbreak Fencing are provided by the reinforcing fibres, environmental properties are largely determined by the polymeric sheath material. Black polyethylene is used as the standard sheath material as it has good all round properties, including excellent resistance to UV degradation, weathering and environmental pollutants. For specific applications, improved flex resistance can be achieved by using an EVA copolymer sheath. Flame-retardant polyethylene-based polymers can also be used.

The fixing point between any composite material and its supporting structure is often a critical point at which tensile or fatigue failure is initiated. The Paraweb Windbreak Fence system uses a simple but efficient channel clamp arrangement which transfers load to the webs but does not produce an unacceptable concentration of stress. The channel clamps firmly grip the individual Paraweb Windbreak Fence webs and allow tension to be applied uniformly, Fig. 1. The result is a durable interface with the supporting structure which has a long service life.

Paraweb Windbreak Fences have been in use for over 30 years and have proved to be extremely durable. Early fences had a web breaking strength of 165 kg and were used to fabricate horticultural windbreak fence structures which were typically 1 to 3 m in height. With the development of industrial and civil engineering applications the strength and stiffness of webs have been increased to reduce fence bow under wind loading and to allow the use of wider post spacing. Table 3 characterises typical Paraweb Windbreak Fence products.

Table 3. Characteristics of typical Paraweb Windbreak Fence products.

Fence type	Core fibre type	Web breaking load (kg)	Fence breaking load (MN/m width)	Fence weight (kg/m^2)
Type 5 (A) 165	Polyester	165	20	0.37
Type 5 (A) 400	Polyester	400	40	0.50
Type 5 (A) 1000	Polyester	1000	100	0.95
Type 5 (A) 1700	Polyester	1700	170	1.04
Type 5 (F) 1000	Kevlar® 29	1000	100	0.55
Type 5 (F) 2000	Kevlar® 29	2000	200	0.60
Type 5 (F) 3000	Kevlar® 29	3000	300	0.75

Fig. 1. Extruded aluminium end clamp for Paraweb Windbreak Fence. The clamp can be seen gripping the individual Paraweb webs. Clamps supplied by Biostructures [7].

DESIGN PRINCIPLES

Designing simple windbreak fence systems is largely a matter of common sense; the following basic principles are applicable to any type of system [5]. Where large or complex fence structures are being designed, input from wind engineers and from wind tunnel modelling is often desirable.

The simplest situation is the one in which the wind strikes normally to the plane of the windbreak fence. In this case a shelter region of 10 to 15 times the height of the fence can be assumed to the lee of the fence. End effects, and the effect of changing wind direction, can usually be accounted for by allowing an inset of about 30° to the shelter region.

If gaps between fences are required, for example to allow exit or entry, overlaps are required and adequate account must be taken of end effects. Failure to do this may result in the formation of a wind funnel.

If protection is required for a track or other linear feature with its axis along the direction of the prevailing wind a series of herring-bone pattern wind break fences can be used.

PARAWEB WINDBREAK FENCE CASE HISTORIES

The following examples illustrate the use of windbreak fence structures to improve sporting facilities in small and large arenas. In both cases the primary objective was to control conditions in the playing area.

Watchorn Tennis Club

The Watchorn Tennis Club, near Chesterfield, UK had a problem similar to that faced by many tennis clubs. High wind speeds across courts were restricting their use and spoiling play for both experienced and novice players.

When the site was developed in the late 1970s it had little natural cover. A 3 m-high Paraweb Windbreak Fence was therefore constructed around the perimeter of the main playing area, Fig. 2. The fence was supported on plain oak posts and was secured using simple wooden battens to trap the individual Paraweb webs. The club manager, Keith Reynolds, noted an immediate improvement in playing conditions, to the benefit of players and the popularity of the club.

Development of the Club has continued and the playing area has been enclosed using an inflated dome textile structure. The Paraweb Windbreak Fence has not been removed as it provides protection to the dome, reducing the effect of distracting wall movements.

Some 20 years after installation, the fence continues to give effective service.

Fig. 2. General view of the Paraweb Windbreak Fence at Watchorn Tennis Club.

Narbonne Athletics Stadium

The decision had been made to redevelop the Narbonne Athletics Stadium (Fig. 3) in the Lanquedoc Roussillon region of France to provide a world class arena. Prevailing winds are experienced from the south (onshore) and from the north of the stadium. The onshore winds were of a relatively low speed and were not considered a problem. The higher speed offshore winds, which had characteristic gusts of about 7 seconds duration with speeds ranging from 2 to 7 m/s, were the subject of a statistical analysis [6] and were considered to be a serious threat. The orientation of the stadium with respect to the prevailing wind pattern is shown in Fig. 4.

Fig. 3. General view of the Narbonne Athletics Stadium showing the Paraweb Windbreak Fence on the right hand side.

A programme of work was initiated to identify the optimum solution for controlling wind speed in the competition area. The work involved the detailed examination of the response of a 1/200th scale model of the stadium complex using the CSTB atmospheric wind tunnel [6]. In addition to predicting the best location for stands and windbreak fences, the work also identified the optimum porosity and fence geometry for controlling short-term gusting.

The options for stadium layout which were evaluated are shown in Figs 4 (b) to (f). The basic condition is shown in Fig 4 (a) and the influence of the existing stands on the reference wind speed of unity (at a height of 10 m) are shown along the front of the northern stand and in the southern corner of the arena.

Fig. 4 (b) considers positioning additional stands to the north of the stadium. Wind speed is reduced by only a small amount. The proposed stands have little effect and are acting as if they were a simple solid wall.

In Figs 4 (c) and (d), the effects of raising the stands which lie in the direct path of the prevailing wind by 3 m is evaluated. An open gap, Fig 4 (c), increases wind speed at the

(a) General arrangement of the stadium
     showing wind speed relevant to
     a reference of unity.

(b) Influence of solid stands to
     the north of the stadium.

(c) Stands 1 and 2 raised 3 m
     from the ground.

(d) Stands 1 and 2 raised and
     with a windbreak fence of
     50% porosity (dotted line)
     in position.

(e) Stands positioned to the south
     and a single windbreak wall
     to the north.

(f) Double windbreak wall
     to the north.

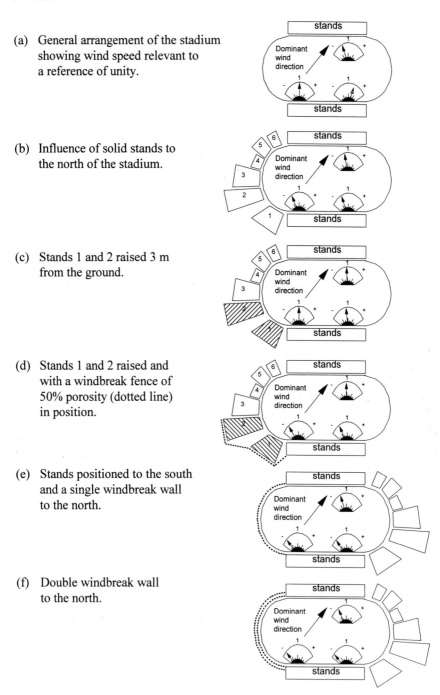

Fig. 4. Summary sketches of wind tunnel evaluations carried out by CSTB [6].
Gauges indicate the effect of each condition on the wind speeds measured on the 1/200th scale model
with a reference wind speed of unity. Measurements were made at a scale height equivalent to 10 m.

reference points; it may also introduce an element of jetting. Filling the gap with a continuous windbreak fence, Fig. 4 (d), significantly improves the situation close to the fence but has no effect on the most distant point across the arena.

Repositioning the stands at the south side of the stadium and introducing single and double windbreak walls of 15 m height along the northern boundary, Figs 4 (e) and (f), produces the most significant improvement at all reference points.

Although providing the most effective solution, the double windbreak wall is also the highest cost option. Advantage was therefore taken of the directional nature of the prevailing wind and the outer windbreak was reduced in size, as shown in Fig. 3.

Having identified an optimum solution for controlling overall wind speed a more detailed analysis of the influence of fence geometry and porosity on gust attenuation was undertaken. Wind tunnel trials indicated that gust attenuation was influenced by:

- distance between the inner and outer fences
- porosity of the windbreak fence panels
- geometry of the outer windbreak panel.

To maintain overall efficiency, the porosity and geometry of the inner Paraweb Windbreak Fence wall was unchanged from the standard configuration of 50% porosity with hole size of 50 × 450 mm. Various alternative geometries and porosities were evaluated for the outer wall leading finally to a 'square' pattern structure which had an overall porosity of almost 60%.

The effectiveness of this design in reducing wind speed and attenuating gusts can be seen in Fig. 5, which shows short-term wind speed with and without the double windbreak fence in position.

The Paraweb Windbreak Fence panels are held in place by reinforced concrete posts which are supported on block foundations. The individual fence panels are held using the Biostructures-designed end clamp, shown in Fig. 1.

Engineering design and erection of the Paraweb Windbreak Fence structure was undertaken by Biostructures srl [7]. The structure has an active area of approximately 2025 m^2 and extends for approximately 135 m with a height of 15 m. It provides effective wind shelter and is favoured by both competitors and the stadium management.

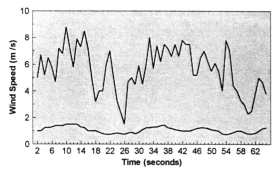

Fig. 5. Wind speeds at Narbonne. The upper line shows typical wind speed data in the Narbonne Stadium region, the lower line shows the equivalent wind speed profile inside the stadium, after installation of the Paraweb Windbreak Fence.

CONCLUSION

Windbreak fence structures are effective in providing shelter for sporting events on both a large and small scale. By careful choice of the position and form of the windbreak fence structure, protection from both steady-state and short-term gust conditions can be achieved.

For straightforward applications, simple design rules can be followed. In more complex applications, input from specialist wind engineers and the use of wind tunnel modelling should be considered. In all cases, adequate consideration should be given to the forces exerted on the windbreaks and adequate post strength and foundation mass provided.

There are several windbreak structures available. Traditional materials suffer in cost terms as relatively closely spaced support structures are required, synthetic windbreaks offer a weight saving but do not allow the use of wider post spacing. The most cost-effective solution is offered by composite synthetic windbreak structures. The high strength composite cladding allows the use of high in-fence tensions and therefore increased post spacing which leads to a reduced overall cost.

REFERENCES

1. Prior and Keeble, Directional wind-chill data for planning sheltered micro-climates around buildings, *Energy and Buildings*, 15-16, 1990/91, pp 997–93.
2. Papesch J G. Wind tunnel test to optimise barrier spacing and porosity to reduce wind damage in horticultural shelter systems, *Journal of Wind Engineering and Industrial Aerodynamics*, 41-44, 1992, pp 2631–42.
3. Sachs P, *Wind Forces in Engineering*, Appendix 4, pp 375–82.
4. Linear Composites Limited, Vale Mills, Oakworth, Keighley BD22 0EB, UK (Fax: +44 (0)1535 643605).
5. Tabler R L, *Snow Fence Guide*, 1991, ISBN 309-05251-3.
6. Gandemer, J, Service Aerodynamique et Environement Climatique, CSTB - Nantes, 11 rue Henri-Pichert, 44300 Nantes, France.
7. G Mougin, Biostructures SA, Cousserans, 46140 Belaye, France (Fax: (+33) 653 62948).

18 SOUND, NOISE AND ACOUSTICS IN STADIUM DESIGN

J. GRIFFITHS
Symonds Travers Morgan, East Grinstead, UK

SUMMARY: The Acoustics Group of Symonds Travers Morgan has worked and is currently working on a number of stadium projects which have required a significant acoustic input at various stages of the planning, design and operation phases of the project. Such projects include the Millennium Stadium Cardiff Arms Park, the Hong Kong Stadium, Wembley Stadium and Ibrox Stadium. Some aspects of these projects are outlined in this paper to demonstrate the various noise and acoustic aspects that need to be addressed when dealing with Stadium projects.

Keywords: Acoustics, construction noise, Hong Kong Stadium, Millennium Stadium, noise impact, noise prediction, planning stage, public address system, sound insulation.

INTRODUCTION

Given the requirement for more stringent safety standards since the publication in January 1990 of Lord Justice Taylor's report on the Hillsborough stadium disaster, there have been numerous proposals to build state-of-the-art stadia in the United Kingdom. As well as the safety requirements, such projects have also been promoted given the increased demand to attend many sporting events in comfort and safety and the desire for stadia to be multi purpose in order to increase the use of the facility and hence boost revenue. Along with all the traditional structural, architectural, quantity surveyor, mechanical and electrical engineering services there has become an increasing need for the services of an acoustic consultant to advise the design team on the inter-related sound, noise and acoustic issues which can have fundamental effects on the planning, design and operational use of a venue.

The acoustic consultant needs to deal with all facets of acoustics from wanted sound to unwanted sound, 'noise'. Aspects related to sound, encompass the quality of the music reproduction and speech intelligibility both of which are functions of the design of the natural acoustics of the stadium interrelated with the design of the sound system and the operational uses of the venue. On the other end of the spectrum, noise also needs to be considered in terms of both the internal and external environment. A sketch illustrating these different issues is shown in Fig. 1.

Stadia, Arenas and Grandstands, edited by P.D. Thompson, J.J.A. Tolloczko and J.N. Clarke.
Published in 1998 by E & FN Spon, 11 New Fetter Lane, London EC4P 4EE, UK.   ISBN: 0 419 24040 3

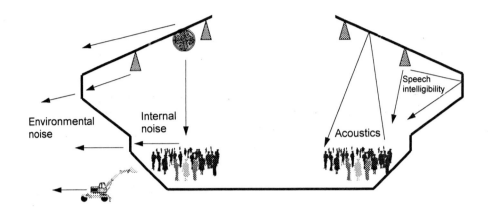

Fig. 1. Sketch illustrating sound, noise and acoustic aspects in stadia.

Millennium Stadium Cardiff Arms Park

The Millennium Stadium is being built on the existing National Stadium of Wales site, with a completion date of Spring 1999. The Stadium is to have a capacity of approximately 75,000 spectators with a retractable roof, the first of its kind in the UK.

Our company was appointed initially by the client to assist in the planning permission stages, where it was necessary to complete a thorough environmental noise impact study of the proposal. More recently, we have been appointed to progress the design of the various acoustic aspects along with the M&E consultants.

For environmental noise impact assessments of stadia, the following noise sources need to be considered:

- construction and demolition
- the range of proposed events (e.g. pop concerts)
- the audience
- public address system
- mechanical and electrical plant
- transportation noise.

Before progressing such a study it is important to obtain a detailed client brief covering all the aforementioned aspects. For example, the client for the Millennium Stadium was keen for the venue to have the potential for increasing its use for pop concerts over that previously permitted (typically two music events per year) whilst improving the environmental impact. For concert activity, the use of the Noise Council's Code of Practice for Environmental Noise Control at Concerts was recommended. For up to 12 events per year the code recommends that the music noise ($L_{Aeq,15\,min}$) does not exceed the background noise (L_{A90}) by more than 15 dB(A) up to 2300 hours. To meet this standard, noise levels would need to be much lower than those which have existed for previous concerts at the old stadium and therefore a design input was required to advise on sound insulation properties and sound control techniques to minimise the sound propagation. Various options are currently under review by the design team which would improve the previous environmental impact from these events. Clearly, in order to assess this and other noise sources, a detailed background noise survey was carried out at various residential and commercial properties surrounding the venue.

Predicting noise levels at a receptor from operational noise sources such as concerts, the audience and public address systems have been developed from data recorded at other venues such as Wembley Stadium. Concert noise source levels were taken to be 100 dB(A) at 40 m from the stage (HSE Research Report 35/1991), whereas audience $L_{Aeq,15 min}$ levels at 5 m have been recorded to be between 88 and 95dB(A) with L_{Amax} levels ranging from 90 to 105 dB(A). The Public Address system source levels were derived from the system achieving a sound level of 95 dB(A) at the audience, which is likely to provide sufficient direct sound when considering the intelligibility of a sound system with respect to reverberant sound and ambient crowd noise (as per BS7827). From the noise source strength of each activity, receptor sound levels can be predicted by accounting for directivity, distance and intervening barriers. A model has been developed which has been validated with the noise levels that were generated by concerts, audiences etc. from the previous stadium during operation.

With the new stadium being enclosed on all sides as compared with the previous venue which had a low uncovered stand to the East, the predicted noise levels are generally lower than those previously experienced irrespective of whether the roof is closed. Reductions of the order of 10 to 15 dB(A) are predicted at properties to the East from all three noise sources. Clearly, the detailed design (building fabric, flanking paths) will have a significant bearing on the sound reduction properties of the stadium as discussed later in the paper.

At the planning stage, it is difficult to be precise about the location and type of plant to be used. A performance specification approach was therefore adopted giving source noise levels at 1 m that should not be exceeded as predicted to meet the advice given in BS4142. The stadium was divided into zones, and source noise levels varying from 65 dB(A) to 82 dB(A) have been specified for various locations around the perimeter, depending on the proximity of nearby residential premises and the prevailing background noise levels. The transportation impact was considered to be small as minimal parking has been made available on the site and there are no significant changes to the present road and rail networks as a result of the development.

Clearly, a significant potential impact, albeit of a temporary nature, is that of construction and demolition noise. A separate report was produced predicting noise levels in accordance with BS5228 for various construction scenarios. As a starting point, the following guidelines (Table 1) were recommended in conjunction with a working code of practice which was provided to assist in negotiating Section 61 agreements under the Control of Pollution Act.

Table 1. Proposed construction noise limits.

Time (T)	Noise criteria ($L_{Aeq,T}$)
0700 – 1900	75
1900 – 2200	65
2200 – 0700	55

The assessment identified a number of impacts, in particular with respect to the first stage of the works which involved piling to construct a retaining wall on the banks of the River Taff. To assess this impact in more detail, four days of test piling were completed and detailed noise and vibration measurements were made. From these tests, an engineering solution coupled with local screening and hours of operation was recommended to minimise the noise impact. Vibration levels from the piling recorded at sensitive receptors were below the criteria recoInmended by BS7385 and BS6472.

Our latest appointment with the M & E consultants, HL&P and their in-house acoustics team has been to deal with the acoustic design of the venue which has involved investigating most of these aspects in further detail as well as considering other design parameters. These include:

- sound system engineering
- environmental acoustics and envelope sound insulation
- internal acoustics and sound insulation between internal areas
- bowl acoustics
- mechanical and electrical plant noise.

The sound system contract has been awarded by the main contractor John Laing plc to Honeywell Control Systems and has been specified to meet all the relevant standards (BS5839, BS7443, BS5588, BS7827). Furthermore, the system has been developed on the basis of a digitally controlled system in order to afford maxiInum flesibility for the various types of events proposed at the stadium. The designs are similar to those our company adopted for Wembley Stadium, which is a digital control and matrix system using fibre optics at the heart of the communication network. The Wembley system, as well as performing the normal safety and event-related announcement functions, has been used by all artists performing at Wembley to supplement their main sound system. This is a unique feature, and is an option that is proposed for the Millennium sound system as it has advantages in terms of both enhanced sound quality in the bowl as well as reducing noise transmission outside the venue. In essence, the sound is distributed from the stadium speakers which focus the sound towards the audience, rather than only using the concert sound system which is normally a single high-powered system which has minimal directivity control characteristics.

In terms of the environmental acoustics, the effect of the building design was investigated against the criteria discussed earlier. The effect of the retractable roof was evaluated as was the sound insulation properties of various building fabrics, in particular in terms of the roof construction. With the proposed roof details for the fixed and retractable elements of the roof, the Noise Council guidelines are likely to be met for most concert events held until 2300 hours (although this is dependent on stage orientation and final design). With some of the design proposals, other events such as boxing may also be able to be held later into the night with minimal environmental impact providing the roof is fully close. Discussions are taking place in respect of the flanking transmission via open vomitories and large ventilation fans/ducts which have the potential of significantly reducing the overall sound insulation. This area of concern is being carefully considered by the design team.

The basic acoustical requirements for the various rooms proposed in the stadium are being designed primarily with respect to their proposed operational use. Acoustic spaces include multi-purpose function rooms, television and radio broadcast suites, executive boxes, plant and equipment rooms, general administration and event-related rooms. Where practicable, optimum Reverberation Times (RTs) have been specified in relation to the size of the room and its intended use. Advice on space planning has been given such that noisy rooms (plant rooms etc) are not located close to acoustically sensitive spaces. Minimum sound insulation properties of 40 dB ($D_{nT,w}$) for general Function Suites to 50 dB ($D_{nT,w}$) for Presentation Suites and 60 dB ($D_{nT,w}$) for Broadcasting Suites have been recommended. Specific guidance has also been given on flanking transmission via ventilation systems etc.

The bowl design is another area requiring careful consideration with respect to acoustics. The design of acceptable acoustics is important in order for the various types of events to be enhanced by achieving good sound clarity for the audience. This aspect is directly linked to the electro-acoustic element of the sound system design so as to ensure a high direct to reverberant ratio thus assisting in providing good speech intelligibility from announcements.

Clearly the internal acoustics are governed primarily by the size, building fabric and shape of the stadium and are indeed variable in nature when considering the effect of the retractable roof. Traditionally RT's have been used to assess the internal acoustic environment, but in large spaces, other aspects such as long delayed reflections and local reverberance need to be carefully examined. At present, acoustic treatment is incorporated within the underside of the majority of the fixed and retractable roof. Further treatment is being recommended for the underside of the middle tier seating and other soffit areas which although having little effect on the overall RT's, would improve the 'local' sound quality for people seated in these areas. An acoustic analysis of the stadium has been completed using computer software with RTs ranging from 10, 4.5 and 3 seconds at low, mid and high frequencies with the roof closed to 7, 4 and 3 seconds respectively with the roof open. Other acoustic parameters are currently being evaluated along with the development of the stadium design.

The aspect of noise from mechanical and electrical plant has been investigated as mentioned earlier for the environment impact stage and further advice is being given as details of the plant become available. The effect of internal noise from ventilation systems is also being considered with reference to the ventilation system design meeting acceptable internal Noise Rating (NR) values for each given room function, including that of the main bowl.

It is hoped that the recommendations with respect to acoustics are fully implemented prior to the construction stage to avoid potential operational problems as highlighted at the Hong Kong Stadium.

Hong Kong Stadium

The need for the services of an acoustic consultant from an early stage of a stadium project and the need for their advice to be followed are demonstrated by our experiences in Hong Kong.

Our commissioning at the Hong Kong Stadium started several months prior to the opening of the stadium in March 1994. The stadium is located on Hong Kong island in So Kon Po and lies at the bottom of a valley with high rise residential dwellings on top of the surrounding high ground. The intervening area comprises of dense vegetation with several villages comprising of temporary housing. Some of these dwellings are located very close to the venue (within 100 m) as is a school to the north west and a hospital to the north east.

The stadium has won several architectural awards and, indeed, provides an impressive landmark, in particular at night during an event. The eastern and western sides of the stadium are covered by a Teflon coated roof which although visually impressive, provides very little sound attenuation. The stadium's operators planned to use the venue for a range of sporting events and concerts. However, our environmental assessment showed that at some of the high rise apartments noise levels from concerts would exceed the limit (65 $L_{Aeq,30min}$) set by the Environmental Protection Department (EPD). High levels far in excess of the EPD limits would also be experienced at the temporary housing. These values were predicted after considering the use of stringent noise control techniques including the use of the stadium sound system with the concert system, the use of field delay towers, careful configuration of the concert system, the construction of a noise barrier, sound propagation tests and continuous noise monitoring and control to comply with the limits set by the sound tests.

As predicted, the noise levels from the first four concert events marking the opening of the stadium, exceeded the EPD limit at a number of properties even though the many stringent noise control methods were adopted. Several canto-pop concerts were then held over the following few months with generally lower external levels (mainly due to lower internal levels acceptable for this type of music) which were still exceeding the EPD limits. Given these breaches, a Noise Abatement Notice was issued by the EPD detailing that all future events are to meet the prescribed limit. Although attempts were made to obtain an exemption for a set number of days, the 65 $L_{Aeq,30min}$ level remained intact and no further concerts were granted.

Further options were considered and detailed tests were carried out on concert systems designed to focus the sound inside the stadium using a number of high directivity devices from the stage and distributed arrays. The stage was also located in a different location orientated away from the majority of the nearest residential premises. The tests showed that the EPD limit could be met at all properties although the seating capacity in the stadium needed to be reduced to just over half the full capacity.

Before agreeing to opt for this solution, a trial concert was staged to assess both whether the audience found the sound level and coverage acceptable and whether the EPD limit could be met in a 'live' situation. The social survey indicated that the majority of the audience were satisfied and indeed the music noise met the EPD limits. However, the noise from the crowd with the music exceeded the limit and a lengthy debate ensured as to whether crowd noise should be included in the assessment. Issuing the audience with gloves was one suggestion!

Having exhausted all acoustic options over a period of a year, the laws of physics with the current stadium design and local circumstances has meant that the situation, to our knowledge has not yet been resolved. This catalogue of events demonstrates the importance of taking acoustic aspects into consideration, alongside all other engineering issues, at the earliest possible stage in the design process. The laws of physics can then be used for, rather than against, us.

19 THE 'SMARTSOUND' NOISE ADAPTIVE PUBLIC ADDRESS CONTROL SYSTEM

B. M. FITZGERALD
Docklands Railway Management Ltd, London, UK
R. M. TRIM
Gilden Research Ltd, Leatherhead, UK

SUMMARY: 'Smartsound' is a new and unique noise adaptive public address (PA) control technology developed in response to a need for the maintenance of a high degree of speech intelligibility against continuously varying levels of background noise and, at the same time, for reduction in unwanted PA noise overspill into neighbouring properties. The system represents a high level of achievement in the successful application of new technological advances in PA electronics, resulting in an effective tool for the control of environmental noise pollution.
Keywords: Communication, environment, noise adaptive, noise reduction, public address, railways, 'Smartsound', speech intelligibility.

INTRODUCTION

Docklands Light Railway (DLR) is a light rail rapid passenger transit system which serves the needs of residents and businesses within the London Docklands regeneration zone. The network threads its way through densely populated areas of east London, often very close to residential façades, so dealing with noise generated by passing trains has been a priority issue for the operator. DLR actually has a policy on noise which states the intentions of the company for the system-wide control of noise arising from train operations and sets down target maximum noise levels at adjacent buildings (see Table 1). In locations where the target levels are exceeded, DLR is committed to using what in law is termed 'best practicable means' to minimise levels of exterior noise at source.

Like all establishments which interface with the general public, DLR depends on good quality public address (PA) for communicating information. Initially DLR stations were generally equipped with a limited number of PA loudspeakers which broadcast messages right down each platform. Variation in volume level was restricted to two settings: daytime (normal level) and night time (reduced level). During the day, PA announcements were often drowned out by excessive background noise, e.g. other trains, traffic, aircraft. At night, the PA system was sometimes perceived as being intrusive by local residents when the railway was being used by only a few passengers waiting on the platforms.

Stadia, Arenas and Grandstands, edited by P.D. Thompson, J.J.A. Tolloczko and J.N. Clarke.
Published in 1998 by E & FN Spon, 11 New Fetter Lane, London EC4P 4EE, UK.   ISBN: 0 419 24040 3

Table 1. Noise and vibration policy target levels.

		Period	Free field Leq dB(A)
Residential areas	Day	07:00-19:00	60
	Evening	19:00-23:00	55
	Night	23:00-07:00	50
Commercial areas	Day	07:00-19:00	60
Schools	Day	07:00-19:00	60

The problem to be solved - complaints about noise
Complaints from local residents regarding noise fall into three categories:

1. Complaints regarding noise from scheduled train services – these are accommodated by the DLR noise policy.
2. Complaints regarding noise from maintenance and construction works (particularly at night) – these are dealt with under the railway's 'permit to work' process which attempts to encourage contractors and railway staff to adopt quieter working methods.
3. Complaints regarding the intrusion of PA announcements (particularly at night) – these are dealt with by the implementation of 'Smartsound' technology.

In response to complaints about the PA system, a survey was undertaken of the acoustic performance of the PA systems installed at every station on the DLR network. The performance was found to vary between stations and sometimes between platforms of a particular station. Inadequate performance was due to reasons such as:

1. Announcements being rendered inaudible by approaching DLR trains.
2. Announcements being rendered inaudible by movements of trains belonging to other railway operators at nearby platforms.
3. Announcements being made at very high levels in order to be heard above nearby traffic noise; these announcements were also, unfortunately, clearly audible in many quieter areas surrounding the station.
4. The normal announcement level contributed significantly to the mainly low levels of background noise around particular stations which was always likely to lead to complaints.

THE PRINCIPLES OF 'SMARTSOUND'

DLR was committed to solving this problem. The objective was to minimise intrusive noise overspill from the station PA system into nearby properties and yet maintain the volume and intelligibility of announcements for passengers. Most of the local residents' complaints about intrusive PA announcements related to problems encountered during the late evening and night when, by definition, ambient noise is lower and people are more sensitive to noise. One solution was to switch off routine messages on the PA system, permitting only emergency announcements. However, this denied passengers on the platforms essential travel information.

A more technically refined solution was required. 'Smartsound' was developed by DLR, jointly with Gilden Research Limited, a specialist electronics company. This new device solves the problem in three ways.

Firstly, by the installation of additional loudspeakers distributed along each platform, each one operating at a lower power output level than the loudspeakers they replaced. Field trials demonstrated a reduction in sound power level of 7.5 dB(A) Leq less than before at a neighbouring property.

Secondly, 'Smartsound' has the capability to adapt automatically the volume of PA announcements in line with the concurrent ambient noise such that the volume level is maintained at a fixed headroom above the constantly changing background noise level. The noise adaptive function of 'Smartsound' is achieved by 'listening' to the level of background noise with a sampling microphone mounted on the station. The 'Smartsound' device detects all PA announcements present and temporary suspension of the 'Smartsound' function prevents the system from suffering from acoustic lock-up.

Lastly, 'Smartsound' uses passive infra-red (PIR) detector switches to activate only those loudspeakers in the vicinity of passengers on the platform. Each loudspeaker is equipped with a PIR sensor switch which detects passengers in its vicinity and so activates the loudspeaker to which it is attached in readiness for the next announcement. Truncated messages can cause confusion and mis-information. The 'Smartsound' PIR device avoids this by sensing the presence of messages, suppressing loudspeaker activation for the duration of that message and then, assuming passengers are still present, permits the next announcement to be broadcast. The device also prolongs loudspeaker activation until the message is finished.

KEY FUNCTIONAL ASPECTS OF 'SMARTSOUND'

The noise adaptive capability of 'Smartsound'

Acoustic noise adaptive public address technology is well established. The core of the 'Smartsound' acoustic signal processor is an electrically controllable attenuator which is connected in series with the PA signal line input, and whose attenuation is controlled by a signal derived from the noise sampling microphone. This line input signal is derived from a Voice Operated Gain Adjustable Device (VOGAD) which standardises the announcement signal level applied to the attenuator for changing levels of line input signal such as is the case for day and night operation and for the varying techniques employed by staff in microphone use. The use of such a simple analogue attenuator enhances system reliability, facilitating setting up at installation and minimising the possibility that settings will be subsequently altered. The whole set up leads to a consistent level of performance.

In using this technique, the 'Smartsound' signal processor dynamically adapts the acoustic level of PA announcements within the areas covered by the loudspeaker system so as to maintain a headroom sufficient to ensure announcements are constantly intelligible above unwanted background noise and yet avoiding adverse environmental effects. Experiments have shown that an acoustic headroom of between 5 and 8 dB(A) is appropriate in most cases.

Sample and hold

The close proximity between microphone and loudspeaker could cause acoustic lock-up, a situation which occurs when the noise sensing microphone receives the station PA announcements and cannot distinguish them from the prevailing ambient noise. As a result it would automatically increase the acoustic level of announcement until upper limit saturation is reached. The station PA announcements would remain at this high level until a sufficient pause in speech occurs to enable the system levels to be re-set. The resulting high acoustic levels of PA announcement would be independent of the actual background noise level and as such would be unacceptable, especially at night. 'Smartsound' resolves this situation with the use of a more sophisticated acoustic isolation technique than is conventionally employed. A specific sample-and-hold technique was developed in order to satisfy such demanding conditions and which resolves the acoustic lock-up effect.

Fail-safe

Within the 'Smartsound' signal processor unit itself, the input line signal is routed through the contacts of a single pole change-over relay which, in the energised state, connects the line input signal to the 'Smartsound' input. A similar relay connects the output of the processor to the station PA amplifier unit. If either or both of these relays are de-energised due to removal of power from the processor, or with failure of one or both of the relays, reversion to non-adaptive operation is automatic which means, in short, that 'Smartsound' is fail safe.

Automatic loudspeaker activation

With reference to the operation of the platform loudspeakers, each is controlled by means of an associated PIR switch unit whose function is to bring the normally quiescent loudspeaker into operation only if the presence of a member of the public is detected. In order to avoid the broadcasting of incomplete PA announcements, each PIR unit is fitted with a circuit which detects the presence of an announcement signal on the line between the PA amplifier and the loudspeaker. Should an announcement be in progress at the same time that a particular PIR unit is activated then the corresponding loudspeaker will not broadcast the remaining portion of that message. Similarly, once a particular loudspeaker has started to broadcast a message, it will not return to its dormant state until the message has been completed.

Two applications have been made jointly by DLR and Gilden Research Limited for UK patents relating to the unique aspects of the 'Smartsound' technology.

Full-scale prototype trials

Islands Gardens station is, and will remain so until the opening of the Lewisham extension, the southern terminus of DLR on the Isle of Dogs and alongside the River Thames opposite Greenwich. The station is elevated with a major distributor road to the north, residential properties to the west, a school to the east and the Thames to the south. Occupants of nearby homes had complained about noise overspill from the station PA system. This location proved to be an ideal site for a full-scale trial of the 'Smartsound' system.

Arrays of additional loudspeakers were installed, generally one per lighting column, in place of the existing two per platform. which broadcast from one end of the platform to the other. Secondly, a single noise sensing microphone was erected on a lighting column at the end of the station closer to the road. Traffic noise was one of the primary sources of excessive background noise. The microphone was connected to the noise adaptive control box located within the station equipment room beneath the platforms. Selected loudspeakers in the open areas of the two platforms were fitted with PIR sensor switches to detect the presence of waiting passengers. Since the commencement of the full scale 'Smartsound' trial in July 1996, the system has performed for 24 hours a day without fault. Noise measurements made have demonstrated a significant reduction in the level of PA announcement volume received at the façade of nearby houses.

Since the evident success of the full-scale trials of 'Smartsound' at Island Gardens, DLR has decided to install the new system at other stations on the network which suffer from less than satisfactory acoustic performance.

THE BENEFITS OF 'SMARTSOUND'

The advantages of 'Smartsound' are numerous. The constant signal to background noise headroom for varying levels of ambient noise results in 'Smartsound' being able to maintain the intelligibility of PA announcements no matter what the level of prevailing background noise is – either high or low. Measurements of speech intelligibility index (RASTI) have shown that values of approximately 0.7 are maintained.

'Smartsound', working in conjunction with distributed arrays of loudspeakers, has been shown to reduce the quantity of noise overspill. Such has been the success of 'Smartsound' in this area that one local resident has been quoted as saying that the quality of her life "...has been transformed".

The operation of 'Smartsound' once set up and commissioned is automatic, giving a constant level of performance. The main controls are tamper-proof as the control panel specifically contains no conventional means of adjustment. The nature of the technology also means that the system reliability is enhanced.

The cost to supply and install the 'Smartsound' system retrospectively at a site already equipped with distributed arrays of loudspeakers, is estimated at merely a fraction of the initial PA system capital investment.

WIDER APPLICATIONS

The use of the 'Smartsound' technique is applicable to any situation where high quality public address announcements need to be communicated clearly to the general public both internally as well as externally, and under conditions of unpredictably changing background noise, most particularly in times of emergency. Such venues include sports stadia, theatres, museums, concert halls and other places of entertainment, plus airport terminals, bus stations and shopping centres. Returning to the transportation field, the 'Smartsound' technology also has application in the field of on-board hi-fi and communications equipment leading to higher quality sound and effective volume control for drivers and passengers alike.

CONCLUSIONS

The 'Smartsound' noise adaptive PA control system was first developed out of a need to resolve specific problems connected either with the inadequate transmission of announcements to railway passengers during periods of intrusive background noise, or with unwanted overspill of PA sound into neighbouring properties. The success of 'Smartsound' has resulted in the development of a new control of environmental noise pollution whilst enabling the user to maintain not only good communication with the public but also customer safety and assurance through intelligible announcements when it matters most. 'Smartsound' also represents a high level of achievement in the successful application of new technological advances in electronics.

20 THE FUTURE OF INTEGRATED ELECTRONIC INFRASTRUCTURE IN MULTI-PURPOSE STADIA AND ARENAS

K. DISS
Philips Projects, Cambridge, UK
I. MAJOR
Philips Lighting, Croydon, UK
S. LIDDLE
Philips Vidiwall

SUMMARY: This paper covers the latest and future technologies drawn from knowledge of current R & D programmes and practical experience gained in projects such as the Amsterdam Arena. You will be given a valuable insight into how an integrated approach for the electronic infrastructure will ensure optimum performance and maximum flexibility.
Keywords: Broadcast TV, CCTV, lighting, smart card systems, sound and communication, Vidiwall.

INTRODUCTION

Modern stadia and arenas face the ever growing challenge of excelling at being the best location for numerous different types of events. Achieving this means maximising on the use of the facility and hence generating increased revenue. Excelling at being the best is not a holy grail, but it is something that careful design and planning at the early stages of conception can have a significant effect upon. An important part of this is the electronic infrastructure and the desired flexibility can only be reached with an integrated system approach in which systems communicate with each other, improving management, reducing operating costs, enhancing safety and ultimately giving more than the sum of the individual components.

SOUND AND COMMUNICATIONS

Sound systems

Sound systems in stadiums and arenas have been rapidly changing over the past few years due to increased demands made upon them and of course, the safety features now required. They have grown from being used to make simple voice announcements and commentary to systems that have to cope with voice, music, alarm and evacuation messages with music varying from the classics to heavy metal rock. This places considerable design challenges in order to provide intelligible sound and full acoustic response, particularly in stadia with moveable roofs and floor surfaces as not only does the input to the system change, but the acoustic response to the environment changes as

Stadia, Arenas and Grandstands, edited by P.D. Thompson, J.J.A. Tolloczko and J.N. Clarke.
Published in 1998 by E & FN Spon, 11 New Fetter Lane, London EC4P 4EE, UK.   ISBN: 0 419 24040 3

well. This in itself is a specialist task in which Philips have considerable experience but the advantage comes in fully integrating the sound system with other electronic systems such as fire alarms and CCTV.

All leading designs are now based around digital processing of the signals and routing of the input to the required areas. This allows far more flexibility and performance from the system, not only improving the quality but also the safety aspect. It has been seen in the past that emergency messages that rely on manual input can cause greater problems than they solve, sometimes because the operator is under stress and gives the wrong information and they can also generate hysteria if the announcement sounds panicked. At other times the message or alarm never gets sent. If however the sound system is integrated fully to the fire alarm and other systems, the appropriate alarm, evacuation message or directions can be automatically triggered and announced in a calm efficient manner with 100% accurate information. Digital processing at the same time automatically ensures the right acoustic profile for the message, changes the volume level to overcome the background noise as monitored locally and prioritises the highest level message accordingly, using simultaneous multi-channel routing as appropriate. All this whilst the system is monitored and performs to the latest safety specifications such as BS 7827, the code of practice for emergency sound systems at sports venues.

Communications

These systems are wide and varied in stadia and arenas, ranging from telephones, mobile radio, paging, intercoms to emergency only systems for fire and evacuation. TV distribution (MATV) can also form part of the communication scene. Again there is the move to digital in these systems and the integration of these systems is important as well. Typically the infrastructure is based on UTP block wiring schemes such as category 5 or 7 linked onto a fibre optic network and it is imperative that all systems can utilise this for maximum flexibility and configuration changes.

The demand today and in the future is for mobile communications and the integration of the communication systems, not only to each other, but also to other systems is an important part of this. For example, it is highly inefficient for a maintenance technician to be called out to the location without knowing what he/she needs to do and therefore probably not bringing with them the right tools. This is solved by direct interfacing between say the BMS system and a paging system so that the technician gets a full report via an alphanumeric pager instead of just a breakdown location.

To cover health and safety aspects, lone worker protection is also called for and the Philips PS6000 system not only gives full alpha-numeric paging, with or without two way speech, but also acts as a automatic alert to a central point if the worker falls over or manually actives an alarm. Naturally, detection of the location of a worker is automatically transmitted at the same time as the alarm and the system constantly self checks against failure.

Further levels of integration are possible with the telephone system so that it can automatically divert to whatever form of mobile communication is chosen, such as GSM, DECT, two speech paging or PMR. This eliminates costly delays in finding the right person, the cost of returned calls and makes for a greater level of staff efficiency.

CLOSED CIRCUIT TELEVISION

Introduction to CCTV in stadia

The larger the number of people at one location, the greater the possibility of a potentially dangerous incident occurring. This is because people often behave quite differently in a crowd than they would alone or in a small group. Added to which, people today are increasingly more stimulated and impatient. In extreme cases, hooliganism and disorder are the results. When a crowd incident does occur, the difficulties of managing it are compounded by the large numbers concerned, and the possibility of unpredictable behaviour en masse.

The use of CCTV has been widely acknowledged as a crucial tool in the management of incidents by providing operators with real-time information of the crowd status. This enables staff to verify incidents and judge the most appropriate course of action.

Crowd surveillance and CCTV

The importance of providing real-time information to the decision makers during crowd incidents has long been recognised. Traditionally, information came from stewards or security staff deployed throughout the area under control, but in recent years CCTV has played an increasingly important role. Compared with direct use of human crowd surveillance alone, cameras offer the following advantages:

- *Speed of response*: following the detection of an incident, when integrated with good communication systems, the appropriate response can be rapidly taken and immediate instructions given based on 'live' pictures in the control room.
- *Cost-effectiveness*: less manpower is required and it can be deployed to respond more effectively to incidents.
- *Safeguarding of personnel*: stewards or security staff are considered as targets by some violently inclined individuals. In such environments, cameras can be used instead of personnel to prevent such attacks.
- *Deterrent effect*: the conspicuous presence of cameras can positively influence the behaviour of a crowd. Potential trouble makers are inhibited because of the increased chances of being identified and caught. Conspicuous camera mounting also gives law-abiding people a feeling of being safe.
- *Consistent quality*: cameras provide a consistent quality of picture. Unlike human observers their performance does not diminish due to being bored, distracted or tired.
- *Analysis:* camera pictures can be recorded on tape. These can then be used to analyse incidents, to effect improvements and to develop new strategies. Tapes can also be used for internal/external training, and handled correctly can assist with criminal prosecutions.
- *Remote control*: remote control functions (pan, tilt, zoom etc.) allow close-ups that would not be possible with the naked eye alone. These functions allow cameras to be optimally positioned or, if required, mounted in concealed positions.
- *Overview*: cameras can be installed in such a way that a total overview can be given of the entire situation (a 'helicopter view').

Application areas for CCTV crowd surveillance in the stadia environment

The use of CCTV for crowd surveillance is absolutely essential in stadia. Even within the defined area of a stadium, the specific tasks the system must perform are diverse. Entrances, exits and turnstiles require monitoring from the point of view of safety to ensure that visitors movement is not inhibited. This ensures that should an incident occur, staff can quickly be deployed to the exact location. The end of an event and emergency evacuations are particularly vulnerable times.

Whether at a sporting event or a concert, the terraces of the stand are areas where incidents can quickly develop. Once again the speedy response of staff is important to minimise the incident and restore order. Other areas inside the stadium that are likely to require security monitoring include security check points, ticket office, retail areas, and betting stations.

Outside the stadium, a particularly vulnerable area is the car park. To attract visitors on a regular basis, the environment must appear secure, and visitors need to be assured that their property is safe. CCTV can assist in the prevention and detection of crime in these areas, and promote the feeling of security amongst visitors.

Characteristics of crowd surveillance pictures
The nature of crowd surveillance pictures can greatly differ from 'normal' CCTV camera pictures used in other CCTV applications. For example, it must take the following into account:

- *Amount of information*: pictures can contain massive amounts of data. To assist the operators it is vital that the equipment, and particularly the camera, is capable of resolving the appropriate level of detail.
- *Variety of information*: the identifying features of individuals must be discernible, such as height, appearance, race, clothing, items being carried etc. The use of colour in the stadium environment is absolutely essential.
- *Movement*: the movement of any individual in a large crowd must be visible. The control given through the use of pan, tilt and zoom cameras is crucial to the systems success.
- *Lighting*: different lighting levels within the field of view must present no problems, e.g. at an outdoor event half the crowd under surveillance could be in bright sunlight and the other half in deep shade. Where the venue is used for concert events, the cameras on the terrace need to deal with near darkness. Furthermore, should stadia lighting fail, operators must not be without the ability to monitor and control evacuation procedures. High sensitivity cameras are as important throughout the stadia terraces and exits as they are externally within the car park areas. The use of infra-red sensitive cameras can be used to good effect.

To accommodate these factors, crowd surveillance applications require cameras that are able to deliver continuous high performance under diverse conditions. The pictures must be of sufficient quality to provide detailed information on which fast, possibly life-saving decisions can be made. Traditionally, monochrome cameras have been used almost exclusively for crowd surveillance applications, but colour cameras bring so much more information. Newer cameras combining both technologies, are also a good choice for certain areas.

The case for colour
Until recently, crowd surveillance systems have generally used monochrome cameras. The benefits of colour coupled with the performance improvements and availability of colour cameras on the market have lead to almost exclusive use of colour in today's stadia. The

case for colour in crowd surveillance is stronger than ever and enables the user to perform a better job in crowd surveillance by providing:

- more information
- better information
- better performance
- more comfort for the user
- more realistic pictures.

One of the most important differences is that the human eye is capable of distinguishing 3000 colour shades against 25 levels of grey of a monochrome camera.

Further technological developments have produce colour cameras that can revert to monochrome in areas of total darkness. When combined with local camera Infra-red lighting, this has the benefit of providing a back-up system should the stadia be plunged into darkness, without losing the obvious benefits of colour during normal working conditions. This technology, such as the Philips 800 series camera, has proved useful covering the terrace areas.

Integration of CCTV
CCTV is just one aspect within the concept of stadia safety and security. Alone, the CCTV system purely provides the information to enable an operator to respond. In crowd surveillance and security applications, the CCTV system will enable the eyes of the operator to be open, but without good communication systems, and incident /alarm systems, the operator is deaf and dumb.

Only through careful integration of CCTV within a total system can the maximum benefit be obtained. The most effective systems are those that alert the operator of an incident in the making; automatically communicate this alert to the management system; automate responses to the alert; and communicate clear instructions to the masses.

This can be simply achieved through the combination and integration of the stadium 'ears', 'eyes' and 'mouth'. For example, the alarm detection systems, such as fire alarms, panic alarms, or stewards with two way radios providing verbal messages act as ears. Similarly, this information can activate an automatic inspection to verify the incident through use of the CCTV system – its eyes. Finally, the operator can either manually trigger a response, or monitor an automatic reaction such as a pre-recorded voice evacuation message or the activation of other alarms. The total responses is designed to guide the stewards to attend to an incident through instructions - using the systems mouth.

All in all, the best systems are those that automate as many processes as possible, to leave the operators free to make judgements. To achieve this, integration of CCTV with other sub-systems is of paramount importance.

INFORMATION DISPLAY SYSTEMS

Display systems for stadia
The purpose of this paper is to identify display systems suitable for use in modern stadia, categorise them into groups according to the type of image required, examine the methods of controlling the image and finally explore the possibility of revenue generation using the display systems.

Image types

Before looking at the different types of displays available it is necessary to identify the different types of information or images that need to be displayed within the modern stadium complex. These different image types can be divided into the following:

Image type	Where used
Action	Main arena, hospitality boxes
Data	Scoreboards
Messaging	Hospitality boxes, entrance halls
Navigational	Entrance halls, concourse
Promotional	All areas
Advertising	All areas

As can be seen, there is a need to display a wide range of images in a number of different geographical locations which often results in a number of different, unconnected and uncoordinated systems being used. The aim here should be to implement a complete image display system for all image types which is controlled centrally.

At this point it is worth considering an important characteristic of the displayed image which has a significant impact on the type of display used and the way in which the display is controlled. The characteristic in question can be best described as the 'update speed' or how quickly and how often a displayed image is updated. The table overleaf shows the differences in 'update speed' requirement for the image types highlighted above.

Image type	Where used	Update speed
Action	Main arena, hospitality boxes	Live (High)
Data	Scoreboards	Medium
Messaging	Hospitality boxes, entrance halls	Low (pre-programmed)
Navigational	Entrance halls, concourse	Low (pre-programmed)
Promotional	All areas	Low (pre-programmed)
Advertising	All areas	Medium

Display types

Today's display systems market contains a wide range of products many of which perform the same or similar functions. This paper discusses products that are believed by Philips to be the most suitable for the applications previously identified.

When choosing a display there are a number of questions that should be addressed to ensure that the display will perform correctly and, most importantly, provide the desired effect. The questions that should be addressed are as follows:

- *Display size*:   How many people need to view the display?
- *Viewing angle*:   Where are the viewers in relation to the display?
- *Brightness*:   What are the ambient light conditions (worst case)?
- *Viewing distance*: How close are the viewers to the display?
- *Display location*: Indoors / outdoors?

Performance specifications vary considerably between display types and therefore care should be taken to ensure that the display is matched correctly to the application.

The table below shows where different types of display technology should be used and the major points for each one.

Image type	Where used	Display type
Action (outdoor)	Main arena	LED
Action (indoor)	Hospitality boxes	CRT Vidiwalls
Data (outdoor)	Scoreboards	LED
Data (indoor)	Scoreboards	CRT Vidiwalls
Messaging	Hospitality boxes, entrance halls	CRT Vidiwalls, Smart monitors
Navigational	Entrance halls, concourse	Smart monitors
Promotional	All areas	CRT Vidiwalls, Smart monitors
Advertising	All areas	CRT Vidiwalls, Smart monitors

LED
: Light emitting diode. Very high brightness. Used for outdoor applications. Available in a range of viewing angles and resolutions. The display is made up from a number of LED modules that fit together to form a single screen.

CRT Vidiwalls
: Cathode ray tube. Vidiwalls are made up from a number of CRT units which are housed in a framework and the image projected onto the screen from the rear.

Smart monitors
: 32", 100Hz, widescreen TV monitor containing a PC with large hard disk and modem / communication capability.

Image control

The most important aspect of any display is the image that is being shown, control of that image is therefore of paramount importance. At any time the image on the display may need to be altered to show for example, a different view of an event from another camera, a new message, a different advertisement, safety or evacuation instructions. Whatever the change to the image is it is necessary that it is controlled seamlessly and that all display devices in the stadium are changed simultaneously. To achieve this level of control PC-based systems are available in two types:

Vidiwall Image Controller (VIC)

The Vidiwall Image Controller is designed to control the images that appear on the Vidiwall. Through the use of VIC multiple images can be shown simultaneously enabling the Vidiwall to convey a number of messages at the same time. VIC is also equipped with a large array of effects to enhance the display and retain the viewers attention these can be pre-programmed into a script and played out as required.

Info display system

The Info Display System is designed for use with the Smart Monitors and allows the program content, the images shown on the screen, to be controlled centrally. Individual images can be created at a central location and sent to each monitor which stores the images on an in-built PC ready for playback at the desired time. Video clips, graphics and presentations can also be stored on the PC and called up and played in any order at any time. The hardware is based on a 32" 100 Hz widescreen TV which is modified to show full VGA and houses a powerful PC. The master screen which contains the PC can also control up to 8 slave units which are also 32" 100 Hz widescreen TVS but do not have the PC built in. All units are able to be switched between data and incoming video and this can also be controlled centrally. The effect of this is that screens for example in hospitality suites, can be programmed to display the event up to a break and then a promotional or advertising message. The Smart Monitor can be programmed to show a different message in each hospitality suite thereby allowing specific targeting of information.

LIGHTING

Stadia lighting

The total lighting budget for a major stadia development accounts for, as a rough average, 1% of the total construction budget; for the Millennium Stadium in Cardiff this would account for a budget lighting cost of approximately £1 million pounds. Unfortunately however lighting is generally one of the first areas that are targeted when cost savings are required, at the end of the day the assumption, rightly or wrongly, is that "a light is a light is a light" and this is all the more true when design and build contracts are used. In such contracts the bottom line price for the project will invariably be the major factor and suppliers and contractors are generally pushed to lower and lower price levels to keep within budget; in many cases the quality of the equipment used is lowered to save costs whilst still meeting the outline specification.

Internal lighting

For the internal lighting within a stadium there is a certain degree of flexibility in the choice of equipment to produce the desired lighting result but it is not always the case that the lowest initial equipment cost provides the best through life operating costs; there are many examples where spending additional moneys on lighting equipment has lead to reduced maintenance and energy costs which have more than repaid the additional investment in under three years. It is all the more relevant now as a result of the 1996 Building Regulations which state that the lighting system shall have an average efficiency of 50 lumens per watt consumed (the lumen being the quantity of light emitted by the light source); with conventional switch start fluorescent systems the notional efficiency is 65 l/w; this can be increased to 75 l/w when using tri-phosphor fluorescent lamps whereas high frequency systems have an efficiency of 99 l/w.

Fig. 1 shows that by using high frequency fluorescent lighting instead of the conventional lower cost switch start systems energy savings of more than 25% can be achieved.

This does not take into account the additional savings that can be made by linking the interior lighting, particularly in office areas to the amount of natural daylight available; is there any need for lighting next to windows to be on in the middle of the day, no; by simply connecting the lighting system to a low cost stand alone lighting control system additional energy savings can be obtained as shown in Table 1.

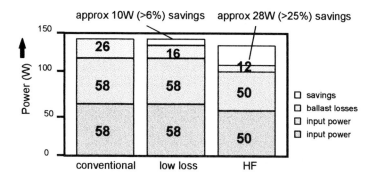

Fig. 1. Input power and watt losses. (Example 2 × 58W TLD.)

Table 1. Daylight linking systems. (HF Regulator and Trios luxsense.)

Indication (%) on energy savings by using daylight adjustments to 500 lux					
Offices	North	South	Corridors	North	South
Summer	45	55	Summer	25	35
Winter	35	45	Winter	15	25

Overall it is possible to save up to 60% of the lighting energy cost by investing more at the beginning of the contract, at the end of the day the client gets a better quality lighting result that saves energy, has lower maintenance costs and is flexible to their needs.

High frequency systems are now commercially available for almost all fluorescent and compact fluorescent lamps, there is no reason, apart from initial extra cost, which in terms of the overall project is minimal, for not using them.

Stadia floodlighting

The same problems that affect the choice of internal lighting equipment under Design and Build contracts also apply to the main stadium floodlighting; unfortunately there is no where near as much flexibility available, the number of users and their individual requirements is much wider than for general internal lighting. Not only must the players be catered for but also the spectators and in large stadia the TV broadcast companies; we all remember the failure of the floodlighting systems at Derby, West Ham and Wimbledon and the dramatic pictures that were relayed around the television networks; the subsequent up-roar lead to the Premier League setting up a panel to investigate the electrical supplies to all premier league grounds; football is now a major business and events such as the three floodlight failures will not be allowed to happen in the future; vast amounts of money have been invested by the television companies and by the traditional loyal fan, into football and they will not tolerate floodlight failures. To ensure it does not happen in the future will require additional investment either at the design stage for a new stadium or as up-grading to existing facilities. Whilst Design and Build is not the direct cause of the failures at the three grounds it will have played a contributory part.

If we examine the trends in sport we find more people are now watching major sporting events such as the World Cup and Olympics either at the events or on the television; this has lead to major sponsorship deals providing additional revenue for investment. As more and more sports are televised the lighting demands of the broadcasters have to be met; as the level of skill on the field increases and the speed of action gets faster then so do the demands on the lighting increase, to see faster action requires higher light levels. The spectators, both at the venue and at home in front of the television want to be able to see the action clearly, to be able to distinguish their team from the opposition and when at the venue feel safe and secure.

Television companies require the optimum lighting conditions to enable them to broadcast excellent quality pictures; they need to be able to do general area shots as well as close ups without losing picture quality. The increased use of slow motion replays also influences the lighting quality, higher light levels are required for high quality slow motion pictures.

The floodlighting system installed has to be able to cope with the demands of these differing groups, never mind if the system also has to be multi-functional and suitable for other sports or events; the considerations are immense and should not be left to chance or compromised in any way. As with the internal lighting additional investment at the start of the project can more often than not yield substantial through life benefits to the end user in terms of maintenance and energy savings.

In Table 2 the main design criteria that should be taken into account are summarised in relation to major stadia projects where international events are envisaged. The requirements are taken from the proposed Sports Lighting European Standard CEN/TC/169N/112E and will form the basis of a European standard.

Table 2. General requirements per segment.

Requirements	Professional segment (International/national)
Suitable for CTV broadcast	Regular use
Colout temperature	4000–6500 K (5600 K preferred)
Colour rendering (R_a)	≥ 65 (90 preferred)
Vertical illuminance level (E)	500–1500 Lux
Horizontal uniformity E_{min}/E_{av}	> 0.7
Spilllight (glare)	Minimal
Investment costs	High

Conclusions
The lighting content of a stadia project is relatively small at approximately 1% of the total budget yet its influence on the final impression of the stadium is immense; people will remember poor lighting; if they feel un-safe they are less likely to return, if the impression of the area is stark and un-welcoming they will be reluctant to return. If the floodlighting is poor the broadcast companies are less likely to come to the venue if there are better grounds close by; the lighting has a tremendous effect on any building not only sports stadia

and it is important that it is not the first area to suffer when savings are required as in the long term this could prove to be false economy.

SMARTCARDS

Smartcards have been around for some time now and are becoming a familiar part of everyday life. However, it is not often realised that there are several types of these cards and they are growing by the day.

Firstly we have the contact type of card which are generally recognised by a small metallic contact pad on one side of the card. These cards have to be inserted into a reader in order for information to be exchanged. The next types are a development of this and are generally referred to as proximity cards where the contacts are not needed and the card can be read by a reader if it is typically within a few centimetres of it. In most cases, these cards are only a one device in that the information on the card is read. The most interesting cards are the long range contactless cards. These cards can be read over greater distances and more importantly can also be written to in order to update the information on the card. The read/write distances are growing all the time and we now have cards that can run multiple applications on a single card. It is these cards that hold the key to the future and Philips are one of the world leaders in the contactless read/write technology, developing not only the cards themselves, but the technology and support software to use them to the limit. These are the areas this paper concentrates on.

The next steps

Imagine the stadia of the future. Imagine the ability to know who was attending the venue, where specifically they were, and what the individuals interests are. This would provide the stadium manager with not only the ability to create a friendly and welcoming environment, but also the ability to generate additional revenue for the stadium through targeted advertising to the individuals tastes.

The Philips contactless smartcard system, 'Smart-Stadia', can bring the future to reality now, and offer both visitors and the stadia managers all manner of benefits..

The system is based on the principle of a visitor card. This could be issued instead of a season ticket to regular users. The card contains an on board microchip that acts as a data store. Data can be written onto the card and read from the card automatically as the visitor wanders through areas of the stadium. Information, such as payment for goods and services can be recorded and debited onto the card, and information such as the individuals seating and purchasing preferences within the stadium can also be recorded and updated. The concept offers the benefit of fast efficient transactions for the visitor.

The benefits to the stadium manager are endless. Personalised services to the visitors can be activated, to maximise expenditure at the stadium. This offers opportunities for loyalty awards and incentive schemes, all of which can be designed to create the desire for the visitor to revisit. Furthermore, with the individuals preference stored on the card, the stadia manager can collect information and store it within a data-mine for market research purposes. Additionally, it can be used to generate revenue for the stadia through individual and targeted advertising. Specific campaigns can be activated as individuals pass a particular retail outlet to dynamically effect their buying patterns there and then.

Other applications for the Smart-Stadia system include the provision for tracking the whereabouts of staff. This would enable stadia managers to quickly re-direct staff in the instance of an incident. Smart-Stadia combined with integrated CCTV surveillance and effective communications can lead to efficiencies throughout the total operation.

The use of help points with seat location plans and multi-language information can also be activated automatically from the card. All of these personal benefits to the visitor are designed to make attendance at the venues a pleasurable experience.

Contact
If you would like to discuss this approach, please contact Christine Spicer at Philips Projects, Cromwell Road, Cambridge, CB1 3HE. (Tel: +44 1223 245191, Fax: +44 1223 413551, email ukzspic@ukccmail.snads.philips.nl).

21 INFRASTRUCTURE REQUIREMENTS FOR SCOREBOARD AND TIMING SYSTEMS IN NEW STADIA

T. E. COLMAN
Omega Electronics UK, Chandlers Ford, UK

SUMMARY: Infrastructure considerations for the installation of large display boards and timing systems are generally overlooked by stadium planners and architects. Size, weight and power requirements for scoreboards are all critical, and for multi-use stadia, the position of ducting for timing system cables and the position of timing control rooms and photofinish equipment complicate the matter still further.
Keywords: Athletics, cable ducts, display technology, photofinish systems, scoreboards, timing equipment, video screens

INTRODUCTION

Architects and planners operate in a minefield of regulations – some clearly defined, and others either imprecise or even totally ambiguous. Building regulations are rarely overlooked, but the actual requirements of the end user may not be fully appreciated. In the case of multi-use stadia the problems may be more severe with different sports (and their governing bodies) having very different requirements particularly regarding scoreboards, prime seating positions and therefore sight lines.

If the stadium is to house athletics competitions as well as team sports (primarily soccer and rugby) not only must the scoreboard location(s) be considered but there is the additional complication of locating timing and photofinish equipment for track events as well as the maze of ducting required for the timing and results systems.

It is impossible to provide all of the solutions to these issues in a single document, but the key pitfalls and infrastructure requirements will be identified below.

DISPLAY BOARDS

Types of display
There are many different 'types' of display (as opposed to display technologies). These include:

- numeric or alphanumeric with individual digits or characters
- full matrix displays – monochrome
- full matrix displays – multi-colour
- video displays (sometimes called 'giant screens').

Stadia, Arenas and Grandstands, edited by P.D. Thompson, J.J.A. Tolloczko and J.N. Clarke.
Published in 1998 by E & FN Spon, 11 New Fetter Lane, London EC4P 4EE, UK. ISBN: 0 419 24040 3

The choice of the type of display is not necessarily simple, and the most expensive type of display (a video display) may not be the best solution for all applications. Certainly video screens are becoming a more or less standard feature of Premier Division stadia, but in many cases these are supplemented by numeric displays (for game time) and even small matrix boards for match statistics and of course the score.

The quantity of alphanumeric information that needs to be displayed for soccer or rugby is quite limited, although team line ups and names of officials may be desirable. It is clear, therefore, that for these team sports the emphasis will be on the provision of video display screens.

However, for athletics, the situation is very different. As for soccer and rugby, there are no specific rules regarding the display requirements, but a substantial quantity of alphanumeric information is required. Generally the solution is to have a display that has two more lines of information than there are lanes on the sprint track, and 32 characters per line. Whilst not impossible, it is difficult to superimpose this quantity of information on a video screen. At the very least, the video display itself would have to be very large to handle such a substantial alphanumeric display requirement.

To complicate matters still further, some video screens do not even have the facility to display text without passing through a complex series of video character generators – a costly and rather unsatisfactory solution.

The 'ultimate' solution for multi-use stadia is to install separate large displays for the video and alphanumeric requirements. This was the preferred solution for the Olympic stadia in Los Angeles, Seoul, Barcelona and Atlanta, as well as some other major venues – particularly in the United States of America. The cost implications, however, are substantial!

Display technologies

There are a number of different display technologies for each type of display detailed above. Initially this may appear to be trivial from the stadium design point of view, but in fact the technology used may have a dramatic effect on the stadium infrastructure. The main technologies – and their advantages and disadvantages from the infrastructure point of view, are summarised in Table 1.

Infrastructure considerations

Apart from the obvious question of running cost, a display with a high power consumption clearly requires a greater installed power. For the largest video screens this may even mean an additional electrical substation (the video display installed in the Waldstadion in Frankfurt – using fluorescent display technology – has an installed power supply of 700 kVA).

The weight, too, is a significant feature. If the display board is to be installed on top of a concrete or steel platform then this may not prove to be a major issue. However if the display is to be suspended from the roof of a grandstand, or even located on top of the roof, then this becomes significant. Once again, it is the major video displays that are the problem – a weight of more than ten tons is possible, even excluding the support structure itself.

Some scoreboards with a low light output may require additional sun screens, or even a completely shaded location, which severely limits the options for the position of the display.

Table 1. Advantages and disadvantages of display technologies.

Electromechanical	Advantages	Low power consumption Good viewing angle
	Disadvantages	Relatively high weight Requires external lighting in low ambient light conditions. High maintenance therefore good access required
Lamp	Advantages	Moderate power consumption Good viewing angle High light output
	Disadvantages	Relatively high weight
LED (light emitting diode)	Advantages	Low power consumption Moderate light output Low weight
	Disadvantages	May have low viewing angle
Video wall	Advantages	Low power consumption
	Disadvantages	Needs substantial sun visor Poor viewing angle Low light output
LCD (liquid crystal display)	Advantages	Low power consumption Low weight
	Disadvantages	May need substantial sun visor Poor viewing angle Low light output
Fluorescent	Advantages	Good viewing angle High light output
	Disadvantages	High power consumption High weight
CRT (cathode ray tube)	Advantages	Good viewing angle High light output
	Disadvantages	High power consumption High weight

Viewing angles should also be considered. Very considerable progress has been made over the past eighteen months regarding the viewing angle of LED based video displays. Nevertheless, some of the lower cost displays, and indeed most LCD units have a limited viewing angle. This means that the spectators sitting more than, say, 50° from the centre line of the display lose contrast, colour balance or even total legibility.

The size of the display is also important. For scoreboards with a high proportion of alphanumeric information one needs to consider the number of lines of information and the number of characters per line. The viewing distance of such text is generally considered to be 500 times the character height. Therefore, for a typical athletics facility with a viewing distance of around 200 m, (with the display mounted at the end of the stadium), the character height will be at least 400 mm.

A video display requires sufficient resolution to give a high quality image – typically no less than 96 pixels – or points of light – in height, and 128 pixels in length. It is unlikely that a video display will be much less than 35 m², so the minimum realistic size for a video board will be around 5 m high by 7 m long. Finding space for such a display may not be easy, particularly when also considering viewing angles and the potential need to replace revenue generating spectator seating by the display board!

Control and peripheral equipment

Having selected the display board, and decided how it can be installed, some consideration must be given to the control system – and its location. Nearly all display boards are controlled from a personal computer – or possibly more than one. However, video displays will generally have a number of additional control devices, generating special effects, interfacing with field cameras – or even external devices – and certainly incorporating preview and editing screens.

This control equipment will generally be air conditioned, and will require a good view of the pitch/track as well as the display board itself. If there is an athletics track, there will also need to be an on-line connection between the scoreboard control system and the photofinish cameras. For this reason, the timing control room and the scoreboards control room should be located either together or at least adjacent to one another.

Major video board installations will include several field cameras, and dedicated cable ducting will be needed from the field area back to the control room.

Data transmission between the control system and the display itself will almost invariably be by fibre optic cable – but again the matter of conduit, or at least a cable route, must not be overlooked.

TIMING SYSTEMS AND PHOTOFINISH

Timing requirements for athletics events are clearly defined within the rules of the sport, particularly regarding photofinish equipment. For any stadium hosting events of regional or superior status, a comprehensive cable network is required serving the whole of the 'infield' area. Further details of these specific cable routes are given below.

With the recent development of digital electronic photofinish systems, it is no longer necessary to have a dedicated photofinish control room directly aligned with the main finish line. However, one or more cameras must be aligned with the leading edge of the finish line and this is not a matter for negotiation! The outfield camera should be located with an inclination of between 23° and 30° to the inside edge of the track. Care must be taken that there is no chance whatever of the camera being obscured by spectators, officials or even suspended TV cables. Television companies themselves consider the photofinish location to be a prime site for their own finish line camera – but under no circumstances can this be permitted. The reasons for this are multi-fold, but there have been numerous instances of television cameramen knocking the photofinish equipment, moving in front of it, or in some cases even disconnecting and moving it! The photofinish cameras must be accessible for alignment purposes, but conversely should not be accessible to unauthorised personnel. Finally, the camera should be securely mounted on a vibration-free platform – suspending from a roof truss is not acceptable as the camera may well be subjected to vibration in windy and other adverse weather conditions.

For major competitions, a second photofinish camera is required, generally mounted on a small tower or strut on the infield area, again directly aligned with the finish line. Cable ducting or conduit will be required from both of these cameras back to the timing control room.

However, not all of the timekeeping activity is restricted to photofinish equipment. Trackside digital displays will be required to indicate running time and the unofficial time of the winner. Wind speed measurement is mandatory for sprint events (as well as several of the field disciplines) and the recorded speed will generally be displayed on more of the digital display units adjacent to the track or runway. Once again, for track events, the anemometer will have an on-line connection to the equipment in the main timing control room.

Photocells will be located at the primary finish line and probably at the two hundred metre point. Start positions will be required at one hundred metre intervals around the track – and will now incorporate acoustic start systems in place of the more traditional start pistols.

Cabling requirements

Some of the specific cable routes have already been identified, and for a stadium only involved in team sports, these requirements are actually minimal. However, the introduction of athletics timing systems into the equation complicates matters considerably. Quite apart from the routes from the field area up to the control room(s), a complex cable system will be installed around the track itself. Connection points will be required for start systems, anemometers, trackside digital displays, false start equipment, electronic distance measurement apparatus, rotating field displays and more.

Waterproof pits will be required at each infield start position, on the outside of the finish line and half way along the main sprint track. Within these pits will be data connection sockets for the equipment detailed above. The pits themselves will generally incorporate a single-phase mains power supply and each pit will be connected to the others by a network of substantial ducts. A minimum of three such ducts are required, one for the mains power cables, one for the timing and results system data cables, and the third for the temporary installation of television camera cables. Each duct should be a minimum of 75 mm in diameter.

It is important that the data and power cables are separated, not only because of potential interference problems, but also to avoid mechanical damage either during initial installation or if the system is upgraded or modified in the future. It is even more important to segregate television cables, which by their very nature are generally installed on a temporary basis, often when a competition has actually started. There have been many well documented instances of television riggers actually disconnecting data cables from timing or results systems to make their own installation work easier!

OVERVIEW OF THE KEY INFRASTRUCTURE ISSUES

It is clear that there are a number of issues that are very easily overlooked. In a short document, it is not possible to list them all – let alone all the solutions. However, an indication of some of the main points to consider is far easier to prepare.

Scoreboard considerations
- Select type of display board, for example video or matrix.
- If the stadium is to host team sports and athletics, will one type of display board meet the different requirements of the two sports?
- Determine the physical characteristics of the board (or boards) that will meet the requirements of the end users (in particular the viewing distances, overall size, weight and power consumption).
- Will the display board be mounted on a fixed platform or be suspended?
- If the display is to be suspended beneath a roof, will the structure actually support the weight of the board itself and the additional steelwork?
- Are the viewing angles and spectator seating areas compatible – or will two display boards be required?
- Identify a suitable location for the scoreboard control room

Timing system considerations
- If the stadium is only used for team sports then this is unlikely to be an issue.
- If the stadium will host athletics events, determine the level of competition envisaged.
- Identify photofinish camera positions at an early stage – they are not negotiable!
- Under no circumstances plan a shared operating position for television camera crew and photofinish cameras.
- Identify a suitable location for the timing control room – preferably with the scoreboard control equipment.
- Ensure that there are suitable storage areas (dry and secure) for the trackside and field equipment, including bulky items such as field display boards, digital trackside displays and false start detection equipment.

Cable infrastructure
- Ensure there are secure cable routes between the control room and the scoreboard(s).
- Ensure that cable 'pits' are planned at an early stage, with adequate drainage.
- Ensure that adequate ducting is installed at field level, with three ducts in all locations.
- Ensure there are secure cable routes, with substantial conduit or ducting, between the field level cable duct network and the timing control room.

22 CONSTRUCTION AND MANAGEMENT OF SPORTS PITCHES IN AN ARENA ENVIRONMENT

J. PERRIS

The Sports Turf Research Institute, Bingley, UK

SUMMARY: Sports pitches in modern stadia are most commonly constructed using the suspended water table principle. With the choice of appropriate materials, recognised playing surface performance criteria can be achieved. In some stadia, environmental effects, particularly low light levels and shade, create environmental stress for the grass. This combined with further stress from increasing usage, particularly of a multi-purpose nature, can cause difficulties for groundstaff trying to present a consistent and uniform surface of adequate turf and playing surface quality. Thoughtful management of stressed areas may help but where stadia design (past or future) creates very difficult environmental circumstances for the grass we are likely to see the development of renewable and/or transportable pitch systems.
Keywords: Environmental effects, grasses, multi-purpose use, photosynthetically active radiation, playing surface standards, rootzone, shade, suspended water table.

INTRODUCTION

The last 20 years have not only seen big changes in the way structures in stadia have been designed and constructed, but also the way in which sports pitches have developed. There has been much research into the science of sports pitch construction and management resulting in well-documented information on materials and design criteria, performance standards and management treatments. Performance and construction standards are readily related to by those in the construction industry, but there is one fundamental difference between sports pitch development and the remainder of stadia construction, namely the importance of biological principles. It is these biological principles which many architects, engineers and constructors fail to fully understand.

The past few years have also seen changes in the way we are using the pitch area within a stadium. Many stadia are now used not solely for one particular sport, e.g. soccer, but for several sports, e.g. soccer, rugby (union and/or league), American football etc., not to say anything of non-sporting events, e.g. pop concerts.

We are at an interesting time in the evolution of stadia design and pitch usage. These days, spectator comfort and facilities are vital but at many venues the stadium development results in significant environmental constraints for the grass and thus playing surface, which invariably is also having to cope with additional stress due to multi-purpose use and extra

Stadia, Arenas and Grandstands, edited by P.D. Thompson, J.J.A. Tolloczko and J.N. Clarke.
Published in 1998 by E & FN Spon, 11 New Fetter Lane, London EC4P 4EE, UK. ISBN: 0 419 24040 3

wear. Poor light and shade as well as poor air movement and temperature extremes are often constraining environmental factors affecting pitch performance.

A good understanding of pitch construction requirements, environmental effects, turf maintenance, together with usage aspects and all their interactions is therefore required by all concerned with stadia development if the turf surface is to sustain appropriate high standards. The sports pitch component of a large stadium development accounts for a small percentage of the total development costs but what price can be put on the importance and value of, say, the Centre Court at Wimbledon during the fortnight of the All England Lawn Tennis Championships?

PLAYING SURFACE STANDARDS

Methods for characterising the playing quality of turf on soccer pitches were developed by Bell and Holmes [1] and standards of playing quality were then developed on the basis of comparison between players' perception of the pitch surface and the results of objective tests taken on the pitches within two hours of matches. Canaway *et al.* [2] have recommended test procedures and test conditions as well as proposed playing surface standards, and these form the basis of current work to produce CEN standards [3] for soccer and rugby.

The attainment of suitable playing surface standards (ground cover, thatch depth, ball bounce, hardness, smoothness, traction and infiltration etc.) is very closely linked with constructional standards and especially drainage and the nature of the rootzone. Furthermore, the choice of appropriate grasses and their subsequent management in the development of a grass cover also has crucial effects on playing surface performance standards.

CONSTRUCTION

A number of methods may be used to build a sports pitch but much will depend upon the money available, standard of sport being played as well as the intensity and nature of usage. At stadia used for professional sport in the UK it is likely that a construction profile based on the suspended water table principle will be built. This method of construction (see Fig. 1) requires a drained subsoil base over which there are layers of carefully chosen materials which will satisfy the drainage stability and agronomic criteria required to produce good playing surface characteristics.

It is also possible to use small grit as the main drainage carpet material as long as certain physical criteria relationships between the small grit and the rootzone are satisfied.

Where a blinding layer profile is formed, one should never consider replacing this intermediate layer with a geotextile membrane; the latter can be prone to eventual "clogging", thus preventing or curtailing water movement through the profile. Geotextile membranes can be used, however, to cover soft subsoil formation surfaces (but not over the drains) prior to introducing the drainage aggregate.

Whilst appropriate drainage and performance from the drainage carpet and intermediate blinding layer is essential, the importance of careful selection of the rootzone cannot be over emphasised. Once selected, rigorous quality assurance procedures must be implemented during rootzone introduction – an increase of just a few percent of fine

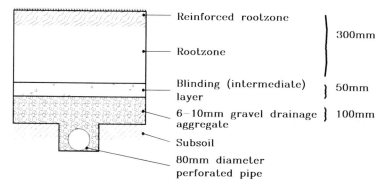

Fig. 1. Suspended water table profile.

particles (clay, silt and very fine sand) can have a major impact on performance, especially drainage. Suitable top quality rootzones for stadia sports pitches are normally acquired from specialist suppliers.

In recent years increasing use has been made of rootzone reinforcing materials in the form of small plastic mesh elements or fine polypropylene fibres. Such reinforcing materials used at appropriate rates can reduce the tendency of the surface to divot as well as increasing stability in situations where the grass is under significant stress due to environmental factors or high levels of usage.

During pitch construction works it will also be importance to install an automatic pop-up irrigation system – so essential in a free-draining construction and where occasional moisture applications will be important as part of agronomic management, not to say anything of water used to try and influence surface characteristics desired by the team manager or coach!

Whilst we continue to use sports pitches in the winter, there is always the risk of matches being cancelled due to frost or snow and in today's commercial world the introduction of an underground soil heating system (if that is the chosen method of pitch protection against frost, etc.) should be undertaken, probably laying the soil heating pipes in the blinding layer of the pitch profile.

Finally, the choice of grasses and method of establishment must be carefully considered. Seeding with *Lolium perenne* (perennial ryegrass) or *Lolium perenne* dominated mixtures is preferred if there is time for good establishment before use. When turfing is necessary, it is worth using specialist turf growers who can supply (given sufficient notice) purpose-produced turf in a rootzone matching that chosen for the pitch. Alternatively, washed turf could be considered but avoid imported turf grown in a rootzone different to that used for forming the pitch.

GRASS PERFORMANCE

Grass (whatever species) must have water, air, suitable temperature, nutrients and light to grow and survive. Certainly the provision of water, air and nutrients should pose no difficulties, and whilst enclosed stadia can influence temperature, the main constraining factor for grass performance is usually light levels. Whilst large stands ensure spectator numbers with good viewing, the invariable reduced light characteristics and significant shading on large parts of the pitch, particularly in the late autumn, winter and early spring period, creates a major stress factor for the grass cover. In low light levels the grass develops thinner, more elongated leaves which have increased succulence and at the same time the sward suffers from reduced density and root growth. The result is often a turf that is weak and more prone to wear and with a greater likelihood of the poorly rooted turf kicking out, i.e. more divots. Reduced air movement in an enclosed stadium also often results in increased damaging turfgrass disease activity with again shaded areas being most prone. We therefore have arrived at a situation in some stadia where, whilst the pitch construction is appropriate for optimum grass and paying surface performance, this is not achievable for significant parts of the year due to environmental effects (particularly low light levels) on the grass. Baker [4] suggests that, where possible, practical steps should be taken to:

- avoid large stands at the southern side
- maintain as much space as possible between the stands and the pitch, especially on the southern side but also where possible for eastern and western stands
- avoid enclosure of corners by covered stands where possible as this affects air movement.

If light is such a constraining factor within many enclosed stadia what can be done to improve this aspect? Sadly at the moment, not a great deal. Whilst translucent roof sheets in the stands may help the situation very slightly, the simple fact is that there is limited knowledge about optimum light levels for sports turfgrasses.

After experiences during the 1994 World Cup Soccer Finals when a natural turf surface was constructed inside the Pontiac Silver Dome in Detroit, Dr J. N. Rogers from Michigan State University felt that in the USA sports turf would require an average of 150 watts per square metre of photosynthetically active radiation (PAR), which is in the 400–700 nm range, for 8 hours on each day. Using these figures, a total daily light energy value would be 4.32 MJ/m^2. Newell [5] feels this figure is on the high side for the UK as daily light integrals only exceeded this figure for three months between August 1995 and May 1996 when PAR was measured. Newell concludes that some 2.5 – 3 MJ/m^2 PAR would be a likely grass requirement for each day. Using Newell's conclusions, Fig. 2 indicates that there is insufficient light for grass during October, November, December, January, February and probably a much longer period on areas in shade. Fig. 3 shows the noticeably poorer light levels on the southern side of an enclosed Stadium and consequently one would expect grass performance to suffer on the southern side.

High lux levels provided by floodlights do not provide the PAR requirements for grass plants and as yet it seems that no commercial lighting organisation is able to practically tackle the PAR requirement by artificial means on a large sports pitch scale. Clearly, further work is needed both to identify theoretical light requirements for sports turfgrasses as well as to hopefully find a practical answer of artificially supplying this requirement.

Where environmental conditions affect grass performance, some adjustment to the turf management may be appropriate.

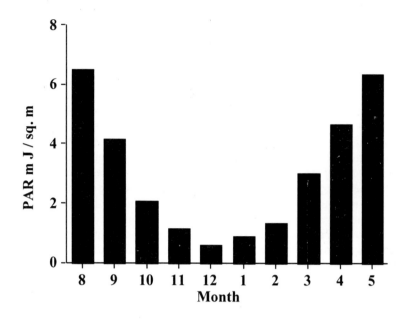

Fig. 2. Average daily values for photosynthetically active radiation available during the winter games season (Bingley, West Yorkshire).

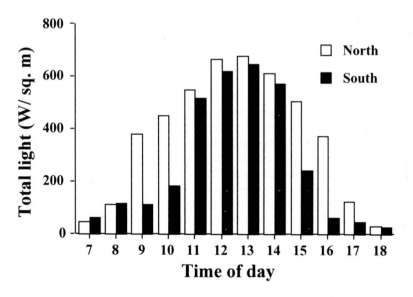

Fig. 3. Hourly light levels on the north and south sides of an enclosed stadium (sunny day in April).

TURF MANAGEMENT

The objective of maintaining a sports pitch is to produce and present a uniform and consistent surface appropriate for the sport being played. When several sports are being played on the pitch, certain aspects of maintenance have to be compromised, for example mowing height when a pitch is being used both for Soccer and Rugby during the winter months. On a well-constructed pitch without any major environmental constraints subject to good management practices it should still nevertheless be possible to present good surfaces for both sports if the surface is not overplayed.

Groundstaff are, however, becoming increasingly concerned about different all the year round use of the sports pitch area. Soccer, rugby union, rugby league and other sports such as American football can result in almost 12 months of the year use with no adequate period to carry out necessary renovation treatments, not to say anything of difficulties in having to implement routine maintenance operations. At sites with less than perfect construction and complementary management it is likely only to be a matter of time before problems with the standard of the playing surface results. The routine maintenance requirements of sports pitches constructed to high specifications are well-documented [6] but it has to be said that the playing surfaces at some stadia never realise their potential because of inadequate skills and sometimes resources. The management of a top specification sports pitch requires an understanding of the science behind the construction as well as the all-important agronomic considerations.

Where environmental effects, particularly low light levels, influence turf and playing surface performance it is worth manipulating traditional maintenance practices. For example, it may be necessary to reduce fertilizer input on regularly shaded areas where there is reduced grass growth. Groundstaff at stadia where there is significant shading often report a reduction of some 25–50% in the amount of grass clippings removed from the heavily-shaded parts of the pitch. Indeed as there is reduced evapotranspiration in shaded areas it is also likely that the sward will require less irrigation. Agronomists at the STRI have certainly reported excessive damage on shaded areas where high fertilizer and irrigation practices, as undertaken on the more open, sunnier aspects of the pitch, have been pursued. To help root development, some careful aeration work at appropriate times should also be practised on the shaded areas, Furthermore, as turf growing in poor light levels is under stress, this should not be further aggravated by excessively close mowing - Soccer pitches should never be mown less than 25 mm and on rugby pitches 40–45 mm.

The sward and consequently playing surface performance is not only affected by choice of grass species, either as a monoculture or in a mixture, but by the performance of the various cultivars of the different species. Turfgrass Seed [7] provides information on various turfgrass cultivar characteristics and should be used to choose a particular grass cultivar that has good performance for relevant characteristics. For example, in stadia where there are regular problems with turfgrass disease activity due to reduced air movement, there may be merit in checking a cultivar's resistance to disease characteristics in addition to other criteria such as ability to withstand wear. Some research work is proceeding at the STRI looking at the ability of some *Lolium perenne* cultivars to cope with reduced light levels and it may well be in time that there could be a positive benefit from this work. Recently there has also been interest in other grass species for use in shaded stadia environments, particular *Deschampsia caespitosa* and *Poa supina*. Further work is needed to conclude on the potential of these particular species.

MULTI-PURPOSE USE

Sports pitches subject to multi-sport and non-sport use the whole year through in a stadium environment where there are significant environmental factors constraining grass performance will probably eventually cause concern. The prospect of developing grasses that can perform better at lower light levels and withstand high levels of wear is possible but I suspect this is a number of years away, if in fact it materialises at all. Some architects may design future stadia more sympathetically with regard to the grasses light requirements and this perhaps is the main hope of agronomists. In existing stadia with continual and heavy demands on the pitch in difficult environmental circumstances there is likely to be an on-going future need for partial or complete returfing of heavily worn and damaged areas. The use of large turf tiles" is already practised at some stadia but there are other possibilities for the future. The use of interlocking small tray modules containing appropriate and complementary rootzone and turf may be a means of repairing areas which need regular renovation. Indeed, future stadia may possibly have parts or the whole pitch constructed with a matrix of modules to facilitate not only regular renovation of wear-prone areas but also the possible temporary removal of part or the whole pitch to leave a surface upon which non-sporting events can take place. Should future pitch developments proceed using such a tray module construction approach then the initial base development of the pitch area has to be reconsidered – probably a drained tarmacadam base would be more appropriate on which to place tray modules. There is already significant interest in this tray module approach – the New York Giants stadium has already successfully introduced and removed and re-introduced their complete playing surface on several occasions with this system. Transportable pitches in small tray modular form or in larger trays could possibly be the direction that future sports pitch development may take at stadia where there is intensive and varied use or environmental problems for the grass. Already a transportable pitch has been developed for use in the Dome stadium in Arnhem, Holland. It is intended to keep the pitch on an adjacent car park area so that the grass can be maintained in good environmental conditions. The pitch will then be moved into the stadium for sporting activity. In such scenarios it is possible, when the pitch is kept outside the stadium, to use the internal area for other activities, hockey on an artificial surface or even flooding the central concrete base area and freezing it to form an artificial surface upon which to play ice hockey. With transportable tray construction systems it will remain essential that the profile build-up within the tray meets the same criteria as identified for the suspended water table construction.

CONCLUSION

Adequate technical knowledge prevails to construct and maintain a sports pitch for significant multi-purpose use in a stadium environment. Crucial for success will be adequate environmental conditions for the grass cover, particularly in relation to light levels. The use of rootzone reinforcement materials combined with thoughtful maintenance may be able to sustain a surface with suitable playing characteristics in many instances but where very heavy usage in difficult environmental circumstances prevails it is likely that we will see the development of transportable pitches either in partial or complete form.

REFERENCES

1. Bell, M. J. and Holmes, G. (1988). The playing quality of associated football pitches. *J. Sports Turf Research Institute,* 64, pp. 9–47.
2. Canaway, P. M., Bell, M. J. Holmes, G. and Baker, S. W. (1990). Standards for the playing quality of natural turf for association football. *Natural and Artificial Playing Fields: Characteristics and Safety features.* ASTM STP 1073 (Eds. R. C. Schmidt, E. F. Hoerner, E. M. Milner and C. A. Morehouse), American Society for Testing and Materials, Philadelphia, USA, pp. 29–47.
3. CEN Turfgrass Committee 217/WG3/N180.
4. Baker, S. W. (1995) *The effects of shade and changes in microclimate on the quality of turf at professional soccer clubs. II Pitch survey. J Sports Turf Research Institute,* 71, pp 75–83.
5. Newell, A. (1997) I Have seen the future and it doesn't work. *International Turfgrass Bulletin.* pp 5,6.
6. *Winter Games Pitches.* The Sports Turf Research Institute. ISBN 1-873431-03-1.
7. *Turfgrass Seed 1998.* The Sports Turf Research Institute.

23 HIGH PERFORMANCE NATURAL TURF SPORTSFIELDS TO MEET THE DEMANDS OF 21st CENTURY STADIA

T. L. H. OLIVER
Tensar International, Blackburn, UK
B. G. CASIMATY
StrathAyr Turf Systems Pty Ltd, Melbourne, Australia

SUMMARY: The paper describes the reasons for recent innovations in sportsfield construction. In particular it deals with a unique system incorporating interlocking mesh element inclusions to stabilise and enhance a sportsfield rootzone. The research background is described together with the development of specifications for a complete system approach for a high use, multi-purpose stadium field. A surface repair method using large blocks of turf and rootzone is described and the extension of this technique for new construction to reduce programme constraints is proposed.
Keywords: mesh elements, multi-purpose sportsfields, rootzone, sportsfield construction.

INTRODUCTION

Natural turf has provided a safe, aesthetically pleasing and low cost surface for soccer, rugby and other outdoor sports since those sports began. Apart from the possibly unique exception of field hockey, natural turf has remained the surface of choice for players and spectators.

Improvements in field constructions, turf maintenance practices, developments in maintenance equipment and new grass cultivars have all contributed to a gradual improvement in field quality over the years. This gradual and incremental improvement has been adequate until recently but the financial demands of the new generation of stadia requires a major leap forward in the performance levels achievable from natural turf fields.

The last few years have seen the introduction of a number of innovative approaches: vacuum drainage systems; movable segmental fields; inflatable field covers; synthetic rootzone inclusions.

The development in the UK of an engineered rootzone system incorporating unique polymer mesh element reinforcement together with the introduction of specialist installation equipment from Australia have been combined with a scientific appreciation and a practical application of the hydraulics of soils to produce a natural turf system capable of delivering the high use and load bearing requirements of the modern multi-use stadium field.

Stadia, Arenas and Grandstands, edited by P.D. Thompson, J.J.A. Tolloczko and J.N. Clarke.
Published in 1998 by E & FN Spon, 11 New Fetter Lane, London EC4P 4EE, UK. ISBN: 0 419 24040 3

This paper presents an overview of the background research together with the design and specification demands of this innovative system and highlights the flexibility in use that it offers to the stadium operator and the construction programming benefits afforded to the stadium architect.

MODERN SPORTSFIELD REQUIREMENTS

A field for a modern sports stadium must fulfil certain essential criteria:

- The playing surface must meet international requirements for the sport to be played in terms of playing characteristics such as ball bounce, traction uniformity, trueness of level etc.
- The drainage of the playing surface must be sufficiently rapid to ensure that the field remains playable at all times. TV schedules do not allow for delayed starts and cancellations are not acceptable.
- The playing surface must offer a safe, consistent surface suitable for contact sports. Injuries are expensive for players and their teams and, in future, litigation against stadium operators may result from poor surface preparation.

In addition to the above sports related properties, the surface should enable multiple use of the facility for concerts and other events in order to maximise revenue on the high capital investment.

The surface should ideally possess the following additional characteristics:

- Be capable of resisting soil compaction to avoid long term maintenance problems following vehicle or pedestrian use.
- Have adequate load bearing capacity to allow occasional vehicle use without rutting.
- Facilitate rapid repair or recovery from localised damage.

THE LIMITATIONS OF CONVENTIONAL SOIL ROOTZONES

Fine textured soils containing significant amounts of clay and silt are prone to compaction and poor drainage capability. They can provide reasonably good surface stability for sports usage if adequately drained, even after significant loss of grass cover. However, their load bearing capability is low and they would deform into a permanent rut if subjected to concentrated wheel loading.

The increased usage rates demanded of stadia sportsfields has led to the development of a number of construction specifications based on high sand content rootzones including in some cases 100% sand rootzones.

High sand content rootzones, if properly constructed and maintained, minimise soil compaction and allow rapid drainage of excess water. The result is a favourable environment for turfgrass roots combined with excellent playing conditions. However, high sand content rootzones, and in particular 100% sand content rootzones, exhibit poor surface stability when grass cover is reduced, as occurs in a soccer goal mouth or high use areas of rugby fields.

High sand specifications such as the USGA Greens Specification have performed extremely well on putting greens for over 30 years. More recently, these specifications have been utilised for sportsfields. However, these sports are significantly more aggressive to the turf surface with higher shear forces and divoting. In order to improve surface stability, specifiers have tended to modify these high sand specifications by shifting to finer textured sands and, in some cases, by reintroducing a limited clay content. The results have been higher stability at the expense of drainage capacity, compaction resistance, root depth and turfgrass health.

Thus, the sportsfield designer is constantly faced with the dilemma of achieving balance between surface stability on the one hand and drainage capability and soil compaction resistance on the other.

THE MESH ELEMENT INCLUSION CONCEPT

A soil stabilisation technique using dimensionally stable mesh elements which does not adversely affect the drainage capacity and actually improves resistance to soil compaction has been shown to shift favourably the relationship between stability, drainage capability (permeability) and compaction resistance.

The mesh elements consist of high strength, discrete 50 mm × 100 mm rectangular pieces of polypropylene mesh with apertures of approximately 10 mm square. The mesh elements are blended randomly throughout the rootzone medium to produce a three dimensional matrix of a relatively fixed but micro flexible nature.

The engineering properties of soils and sand blends incorporating various concentrations of mesh elements have been thoroughly investigated. The mesh elements have been proven to increase bearing capacity (Table 1), increase shear resistance and beneficially alter the nature of a footing type failure without adversely affecting the surface stiffness (resilience). (McGown *et al.*, 1985; Qayyum, 1995; McGown *et al.*, 1997).

Eight key studies have been conducted since 1985 at Texas A&M University including five long-term field investigations, each of four-plus years' duration. (Beard and Sifers, 1989, 1990, 1993; Sifers & Beard, 1997; Sifers *et al.* 1993, 1996). In addition, there have been two separate studies at the Sports Turf Research Institute, Bingley (Canaway, 1994; Richards 1994).

Table 1. Bearing capacity improvement ratio (I_f) for two rootzone profiles with perennial ryegrass turf cover, under vertical and inclined eccentric loading conditions, at a relative settlement of 10% of footing width.

Profile	Vertical loading	Inclined (20°), eccentric loading
USGA without mesh elements	1.00	1.05
USGA with ReFlex mesh elements (0.44% by weight)	1.32	1.41

(after Qayyum, 1995)

Table 2. Effects of ReFlex mesh elements on surface hardness of four turfed rootzones.

Rootzone texture	Clegg Impact Value (g), 2.25 kg hammer, fourth drop from 300 mm height		
	No mesh	Mesh	% change
High sand	88	69	-19
Sandy loam	97	76	-19
Sandy clay loam	107	84	-23
Clay loam	116	87	-29

Note: Values are means of multiple observations over three-year observation period (after Sifers & Beard, 1996)

The findings revealed several major agronomic benefits of major significance for sportsfields.

- reduced surface hardness (Table 2)
- improved soil water percolation and infiltration rates (Table 3)
- improved air filled porosity
- a resistance to soil compaction
- reduced divot size (Table 4)
- enhanced divot recovery rate (Table 5)
- improved uniformity of ball bounce.

The improved soil water percolation, improved air filled porosity and compaction resistance have been ascribed to a unique feature of the mesh elements – their ability to flex within the soil structure. It is hypothesised (Beard and Sifers, 1993) that when the rootzone is compressed by traffic at the surface, the mesh elements flex then recover on removal of the load to produce a microcultivation type action within the soil. This helps keep the soil structure open, improving both percolation and aeration of the rootzone.

Table 3. Comparisons of water infiltration into mature turfed rootzones modified by 0.44% by weight ReFlex mesh elements versus not modified.

Rootzone texture	Infiltration rate (mm/h)		
	No mesh	Mesh	% change
Sand	571*	1069*	+47
Sandy clay loam	<10	113	+91
Clay loam	<5	75	+93

* Means of four individual assessments per year over three years (after Sifers and Beard, 1996)

Table 4. Comparisons of divot opening length and width for mature Tifway bermudagrass grown on three distinct rootzones modified with 0.44% by weight ReFlex mesh element inclusions versus not modified.

Rootzone texture	Divot opening length (mm)			Divot opening width (mm)		
	No mesh	Mesh	% change	No mesh	Mesh	% change
Sand	134*	102*	-24	55*	46*	-16
Sandy clay loam	141	95	-33	49	42	-14
Clay loam	149	76	-49	54	42	-22

* Means of four individual assessments per year over three years (after Sifers and Beard, 1996)

THE STRATHAYR SYSTEM PROFILE

The benefits of the interlocking mesh elements have been proven for a wide range of soil types. However, in order to optimise performance, for stadia sportsfields a high sand content rootzone is preferred.

The StrathAyr construction profile utilises the perched water table concept.

This design concept comprises a carefully selected sand rootzone overlying a compatible gravel base. Provided that certain particle size distribution rules are rigorously observed, the abrupt change in particle size from rootzone to gravel layer enables moisture to be retained (perched) in the sand rootzone owing to capillary suction.

The StrathAyr system profile incorporates mesh elements and an organic amendment into the upper layer of the rootzone.

The depth of the rootzone layer is critical and is based upon the **measured** hydraulic and capillary suction properties of the selected rootzone and gravel materials.

Drainage pipe sizing and spacing are calculated to suit the rainfall intensities at the intended location. The layer thicknesses and the uniformity of the profile play a crucial role in the final field performance. For this reason, specialist equipment has been developed to lay the various layers to the high tolerances required

Table 5. Comparisons of divot opening turf recovery time for mature Tifway bermudagrass grown on three distinct rootzones modified with 0.44% by weight ReFlex mesh element inclusions versus not modified.

Rootzone texture	Divot opening recovery					
	Days to 50% recovery			Days to 75% recovery		
	No mesh	Mesh	% change	No mesh	Mesh	% change
Sand	21*	14*	-33	28*	20*	-29
Sandy clay loam	32	19	-41	41	26	-37
Clay loam	35	25	-29	46	30	-35

* Means of four individual assessments per year over three years (after Sifers and Beard, 1996)

PERFORMANCE SPECIFICATIONS

With the introduction of a number of competing sportsfield 'systems' and the various design approaches adopted by specialist consultants, there is a need to move towards specifying sportsfields by **performance criteria**. Pre-defined performance criteria are the only sensible means of assessing bids for competing systems.

A performance specification should cover the most important performance criteria. An objective measurement approach can replace a subjective and unscientific decision making process.

- installation/percolation rates
- surface hardness
- surface traction
- ball bounce height/consistency
- load bearing capacity (for multi-use fields)
- surface level tolerances.

A typical performance specification for the StrathAyr system is shown in Table 6.

Table 6. Typical performance specification for StrathAyr system sportsfield (StrathAyr, 1996).

1	Field infiltration rate	Rootzone infiltration rate (with 10 mm max thatch cover) to be greater than 100 mm/h after compaction of 80 j/m^3 using double ring field infiltrometer.
2	Drainage capacity	Effective drainage capacity that is capable of removing an infiltration of 100 mm/h over the entire field area.
3	Profile integrity	Having no significant migration of rootzone into gravel layer.
4	Moisture retention	Rootzone to have a moisture retention over the entire field between 10% – 20% by volume after compaction of 80 j/m^3 within the top 50 mm of the rootzone. Measured on samples 50 mm dia, 50 mm depth with thatch removed using oven dry method. Samples to be taken 2 – 20 hours after rainfall or irrigation period during winter.
5	Field porosity	Rootzone total porosity to be between 36% – 50% after compaction of 80 j/m^3 in the top 50 mm of the rootzone but below thatch level.
6	Surface hardness	Hardness index readings to be between 50 – 90 rating from 0.5 kg Clegg hammer dropped from 300 mm height when applied to the full range of moisture throughout the year.
7	Surface structural performance – modulus of reaction	a) Minimum modulus of 9.0 x 10^4 kN/m^3 at surface deformation of 3 mm. Measured by plate bearing test (305 mm dia plate) [B.S.1377:Part 9: 1990, Method 4.4] b) Initial recoverable deformation minimum 15%

REPAIR TECHNIQUES

Experienced groundstaff have always used techniques whereby small plugs of turf are transplanted to repair deep divots or damaged turf. This concept may now be extended to cover large areas of the field or even an entire field.

Large reinforced blocks (up to 2.5 m × 2.5 m) of the upper rootzone layer, typically 100 – 150 mm thickness (known as StrathAyr Squ*Ayr*s) can be pre-grown off site and transplanted into the field to replace entire worn or damaged areas.

The mesh element inclusions in the rootzone stabilise the blocks and enable the large blocks to be handled by overhead grab, simplifying the handling process.

This technique has been successfully used to repair soccer goal mouths, replace large areas painted with logos and even to transplant an entire cricket wicket. The method has proven successful at a number of prestigious stadia including Old Trafford, Manchester; MCG, Melbourne and Parramatta Stadium, Sydney.

The method is being adapted to enable sections of a field to be swapped around, substituting high wear areas for low wear areas and low light incidence areas with higher light incidence areas of a field.

With the advent of closed stadia and the inherent shading and static air problems that these bring, the ability to cycle areas of the field has many advantages.

The same techniques can be employed for new construction to overcome tight construction programme constraints.

The entire field area may be pre-grown off site allowing the mature turf surface (and upper 100 mm of rootzone) to be transplanted to the stadium and be ready for use immediately. This operation can be carried out virtually any time during the year and is not restricted to the growing season.

SUMMARY

The financial returns achievable from modern stadia are directly influenced by the durability and flexibility in use of the playing surface. It is essential that high tech stadia include high tech fields.

Player safety is a second but no less important consideration which, in turn, has major financial implications for players themselves, their clubs and the stadium operators.

The paper describes in detail one field system that addresses both the high use issue and player safety.

REFERENCES

1. McGown, A., Andrawes, K. Z., Hytiris, N., Mercer, F. B. Soil strengthening using randomly distributed mesh elements. *Proceedings of the 11th International Conference on Soil Mechanics and Foundation Engineering*, San Francisco, 1985.
2. Qayyum, T. I. *Bearing capacity of unreinforced and reinforced soil under rapid loading.* PhD Thesis. University of Strathclyde, Glasgow, 1995.
3. McGown, A., Qayyum, T. I., Beard, J. B., Oliver, T. L. H. Performance evaluation of turfgrass root zone materials and profile constructions using an innovative rapid,

eccentric loading test method. *Proceedings of the International Turfgrass Research Conference*, Sydney, July 1997.

4. Beard, J. B., Sifers, S. I. A randomly oriented, interlocking mesh element matrices system for sport turf rootzone construction. *6th International Turfgrass Research Conference*, Tokyo, 1989. 6, pp. 253–7.

5. Beard, J. B., Sifers. S. I. Feasibility assessment of randomly oriented, interlocking mesh element matrices for turfed rootzones. *Natural and Artificial Playing Fields: Characteristics and Safety Features.* American Society of Testing Materials, Philadelphia, Standard Technical Publication STP1073, pp. 154–165.

6. Beard, J. B., Sifers, S. I. Stabilization and enhancement of sand-modified root zones for high traffic sport turfs with mesh elements. *TAES Research Bulletin.* Texas A&M University, B-1710, February 1993.

7. Sifers, S. I., Beard, J. B. Enhancing participant safety in natural turfgrass surfaces including use of interlocking mesh element matrices. *ASTM Symposium on Safety in Football*, November 1994. ASTM STP1305, pp. 156–63, 1997.

8. Sifers, S. I., Beard, J. B., Hall, M. H. *Turf plant responses and soil characterizations in sandy clay loam and clay loam soil augmented by turf in interlocking mesh elements - 1992.* Texas Turfgrass Research - 1993. TAES PR-5142. pp. 112–6.

9. Sifers, S. I., Beard, J. B., White, R. H., Hall, M. H. *Assessment of plant morphological responses and soil physical characteristics resulting from augmentation of sandy clay loam and clay loam turfgrass root zones with three densities of randomly oriented interlocking mesh elements - 1993.* Texas Turfgrass Research. TAES Turf 96-7, pp. 36–41, 1996.

10. Canaway, P. M. A field trial on isotropic stabilisation of sand root zone for football using Netlon mesh elements. *STRI Journal* 70, 1994.

11. Richards, C. W. The effects of mesh element inclusion on soil physical properties of turfgrass root zones. *STRI Journal* 70, 1994.

PART FIVE

NEW STADIA PROJECTS IN THE UNITED KINGDOM

24 THE MILLENNIUM STADIUM – CARDIFF ARMS PARK

P. D. THOMPSON
Millennium Stadium plc, Cardiff, UK

SUMMARY: A vision was created in early 1995 by the Welsh Rugby Union (WRU) and the County Council in Cardiff to create a new Welsh National Stadium for the Rugby World Cup in 1999. Key ingredients enabled it to happen.

Firstly, the WRU decided that, having spent 15 years from 1967 creating their unique stadium at the Cardiff Arms Park, one of the most famous rugby grounds in the world, and only completed in 1983, they must build a brand new stadium if they were to compete with the rugby nations of the world and go on to the next millennium and provide modern exciting facilities for the people of Wales.

Secondly, the WRU, having ventured into multi-use activities, decided to commit to a stadium with a capacity of 72,500, that would guarantee 'Surety of Event' and decided to incorporate a retractable roof over their hallowed turf, which will be the largest retractable roof stadium in the world and the only one in the United Kingdom.

Thirdly, the Millennium Commission was looking for eight major projects in the UK with national impact for the community, creating major regeneration and with a major event to celebrate the next millennium – Millennium Stadium fitted the bill.

Keywords: Cardiff, City centre, funding, multi-use stadium, project management, redevelopment, retractable roof, rugby.

INTRODUCTION

I have believed from the day I was appointed as Project Coordinator for the new Welsh National Stadium that the vision created in early 1995 by the Welsh Rugby Union (WRU) and the County Council in Cardiff would become a reality for the Rugby World Cup in 1999. The key ingredients necessary to allow this to happen are so exceptional that it is not possible to doubt it.

Cardiff Arms Park is one of the most famous rugby grounds in the world and the National Stadium on this site is owned and controlled by the Welsh Rugby Union. The Welsh Rugby Union is the controlling body for rugby in Wales involving 220 clubs and a further 175 affiliated clubs throughout Wales.

The Stadium is located in the heart of the City of Cardiff, the Capital City of Wales, alongside the River Taff and adjacent to the main rail and bus stations.

In June 1994, the Union set up a ground redevelopment committee to look at the possible redevelopment of their Stadium and joined with Cardiff County Council later in

Stadia, Arenas and Grandstands, edited by P.D. Thompson, J.J.A. Tolloczko and J.N. Clarke.
Published in 1998 by E & FN Spon, 11 New Fetter Lane, London EC4P 4EE, UK.   ISBN: 0 419 24040 3

the year to find ways of linking the redevelopment of the Stadium to the regeneration of the West of the City.

Early in 1995, the opportunity for the Union to bid to host the Rugby World Cup in 1999 became available and, against severe competition from the Southern Hemisphere, they succeeded, setting a critical completion date if the ground was to be redeveloped on its current site. The basis of the World Cup Agreement related to the National Stadium as it was in 1995, not on the basis that it would be redeveloped; whilst the World Cup is a very significant event, any investment in a redeveloped stadium would need to be based on a 50-year life span.

The Union looked at its status with reference to its facilities in comparison with the other Rugby home nations and it was clear that, whilst the Cardiff Stadium was one of the best in the world when it was designed in 1962, it had been long since overtaken. England had just completed their new 75,000 capacity stadium at Twickenham, Scotland their new 67,000 capacity stadium at Murrayfield, and France are constructing their new 80,000 capacity stadium in Paris for the Football World Cup in 1998. Ireland are also proposing a new stadium. In addition, the Union looked at other National Stadia in the UK and identified that Wembley had plans for major investment. It was clear that, without significant investment, Wales, Welsh Rugby and its Capital City would be left behind if action was not taken.

THE EXISTING STADIUM AND THE BRIEF FOR THE NEW STADIUM

The National Stadium Cardiff Arms Park had a capacity reduced down to 53,000 with 11,000 of those standing. The reductions since the Stadium was designed were due to safety requirements and the likely requirement for all-seater stadia would have reduced the capacity even further to 48,000.

The previous stadium was predominantly a rugby stadium for international football and many other leisure and community uses including boxing, pop concerts, bands, religious festivals and choirs. This multi-use requirement created the requirement from day one for the new Stadium to have a roof and to provide a natural grass pitch for rugby. It was decided that a retractable roof should become part of the Union's brief. The only retractable roof stadium developed in Europe is the Amsterdam Arena with a capacity of 50,000.

The key facilities that were required to be provided by the new Stadium were set in stone (or in this case steel and concrete):

- multi-use, third generation, retractable roof stadium
- 75,000 capacity
- hospitality, bars, restaurants, shops, fast food outlets
- offices, entertainment facilities, training and fitness centres
- world rugby experience/roll of honour/visitor attraction centre
- special facilities for families and the disabled.

The hallowed turf of Cardiff Arms Park is famous all around the world and its history goes back to 1874. The first Grandstand was erected in 1885 and the new pavilion servicing rugby and football in 1904. Greyhound racing commenced in 1927 and a new North Stand was built in 1934. In 1958 the site was used for the Commonwealth Games and in 1962 the then-new design for the National Stadium was produced by O. V. Webb. The construction process went on from the North around to the East from 1968 to 1983.

The site, whilst being in the city centre, has always been an open site linked to the river to the West and the castle grounds to the North. Jointly operated for rugby, cricket, hockey and bowls up to 1966, when the last cricket game was played, the site became two rugby grounds, one operated by Cardiff Athletic and the other by the Welsh Rugby Union.

The site was created when the River Taff was straightened in 1851 to facilitate the construction of the new Central Railway Station. The river tributaries and canals operating in the area give clues to its operational uses, with Quay Street opposite the entrance to the Stadium and the City Arms pub adjacent, which gave the name to the Arms Park.

The design of the previous Stadium was magnificent, providing a tight bowl close around the pitch and an atmosphere second to none. However, the requirements of the time were dictated by the need for seats to enjoy the joys of a very successful Welsh Rugby Team, then one of the best in the world. In terms of today's facilities, it had none, and the tightness of the site did not allow for development. All that existed was bare concrete concourses, basic toilet facilities, primarily male, and the odd burger stand alongside the river. The current requirements as laid down in the Green Guide were not developed at that time in a way that has been demanded by a number of tragic stadium disasters.

The way forward was seen to be very different. Firstly, the Development Committee considered covering over the East Terrace, but not only did this reduce capacity but it had serious problems with the conservation area requirements of Westgate Street. The next stage was to consider developing the whole stadium on its existing footprint by adding a third tier and introducing extra facilities. This would have been very expensive but was rejected primarily because there was not enough room on the site and the restrictions of egress would have meant that it would not be possible to obtain a safety certificate for the capacity required.

The team decided to look at facilities that had been developed and operated and this inevitably led them to the USA and Canada:

- The Georgia Dome, Atlanta has a fixed roof and artificial pitch but has very similar criteria to that required by the Welsh Rugby Union.
- Cominski Park in Chicago, where the relationship between food, drink and facilities for all were clearly evident and to understand the relationship of the seating to the pitch.
- The Joe Robbie Stadium, to look at the hospitality facilities, quality of fit-out and a natural pitch.
- Madison Square Garden, as this was the only stadium providing access to the players' facilities, and of course the Sky Dome and its Mega Screen.

The team dreamed of the excitement of new stadia in Japan and came down to reality when it was established that the minimum cost outlay to achieve their goal was well in excess of £100m and many of the Japanese stadia had a cost of £250m.

THE COMPANY AND THEIR PROJECT TEAM

Millennium Stadium Plc was registered in February 1996 and became operational in July 1996 as a Public Limited Company. The Company's purpose is to develop and operate a new multi-use stadium under a 50-year lease from the Welsh Rugby Union.

The control of the company lies with the Welsh Rugby Union with five directors and 50,000 ordinary shares all voting, as against the County Council who have four directors and one special non-voting share. This special share is to give comfort to the Millennium Commission requirements as a non-participating funder, namely charitable and community considerations and matters prejudicial to the project.

To satisfy these requirements, a Charitable Trust was established in July 1996 to support community and sporting charitable activities in Wales. The Charitable Trust will receive from the Stadium Company 10% of distributable profit and a ticket levy of 25p per ticket or 1%, whichever is the greatest. Additionally, the Stadium will be made available 15 days per year free of charge for community events.

THE PROJECT TEAM

The Union and the County Council understood that specialist project management expertise was needed to turn their dream into reality and I was appointed as project co-ordinator in February 1995. The first task was to choose an architect with the appropriate experience for such a specialist design and preferably British if not Welsh for a National Stadium for Wales. Presentations had been received by the WRU from a selected number of designers to put forward proposals for the redesign and Lobb Partnership were chosen as lead designers as architects. Lobb were identified as leading British stadia designers specialising in stadia and arena design, having just completed the McAlpine Stadium at Huddersfield with the RIBA. Design award in 1995, and were involved in many major stadium designs around the world. Our confidence in Lobb was confirmed when they were appointed as designers for the Sydney Stadium for the Olympics in the year 2000. From the selected designers W S Atkins were chosen as engineers to the project, leading designers of exciting projects around the world with a major presence in Wales. At a later date Hoare Lea and Partners were appointed as Services engineers.

It was clear from day one that an experienced project management organisation was needed to deal with the complex interfaces that would be created by the development of a new stadium in a city centre with buildings all around. Experience in major stadia projects was important and O'Brien Kreitzberg was chosen as the foremost construction managers in the USA, and project managers not only with experience in major projects, for example J. F. Kennedy Airport Redevelopment, but involvement in the management of stadium projects including the infrastructure requirements for the Atlanta Olympic Facilities. To ensure that the commercial controls had a local perspective to maximise all the important aspects of the Cardiff location of the development Bute Partnership, quantity surveyors were appointed.

The project team established its basis of control: Project control; cost control; design control; construction control resulting in overall control. Extensive interviewing of contractors in the UK and Europe resulted in John Laing Construction being appointed as main contractors; they took responsibility for the design team and agreed a design-build contract. It had been decided to contract under the Engineering Construction Contract, which encouraged teamwork and recognised the role of the project manager. It was decided to use Option C, which was cost plus having firmed up preliminaries and a management fee. Laing have recently successfully completed the design and construction of the new Severn

Crossing and the project team have every confidence that they will successfully complete the Stadium by June 1999 ready for the World Cup. With such an important deadline in front of them it was imperative that above all else the Stadium is completed on time.

FUNDING

It became apparent that the budget needed for the Stadium was well in excess of £100m and that Government monies would be needed if the project was to succeed. The site was not in a grant-aided location and, unlike most parts of the world, the UK Government does not automatically support national arenas with public money as part of its responsibility. The National Lottery came into being in the UK in 1994 and the only opportunity available to a major project in Wales was the Millennium Commission. The unfair way that monies are distributed for major projects means that, because Wales has a population of 3 million, monies allocated to Sport, Heritage and Art are so small that they are unable to support a major project. This means that the one-off opportunity to obtain monies had to rely on convincing the Millennium Commission to support the Stadium as one of the eight major projects in the UK that they were committed to. If this failed this was the end of the redevelopment of the National Stadium of Wales for the foreseeable future. The criteria laid down by the Commission were to: have public support; make a substantial contribution to the community; look back over the past Millennium and into the new one; mark a significant moment in history; be of high architectural design and environmental quality; include partnership with the local community; and not to be possible without Commission funds or normally be part of the responsibility of public funds.

The new National Stadium met all these criteria. Unfortunately, unlike the other lottery funders, the Millennium Commission are only allowed to provide a maximum of 50% of the cost or £50M. However, the Welsh Rugby Union decided to prepare a bid for £50M based on a likely cost of £114M leaving £64M to be raised from the commercial opportunities created by the new stadium. The project identified the proposed stadia as the Euro Stadium as it would be the most advanced stadium in Europe and would attract events of all types from all over Europe.

The first bid made in March 1995 was based on turning the Stadium 180°, taking over the adjacent Cardiff Ground and building them a new stadium in Cardiff Bay. Linked to this proposal was a new Olympic Pool and Hotel Complex. However, the Millennium Commission in October 1995 decided not to support this proposal, primarily due to political pressure of having at that time an alternative exciting project in Cardiff to consider, the Cardiff Bay Opera House, and only one project could be supported. The official reason given was the unacceptability of supporting two stadia in Cardiff, one of them a club venue.

The project considered the alternatives available and decided that the only way to combat the Commission's objections was to leave the Cardiff Ground intact and develop over the site then occupied by the T.A.V.R.A. and British Telecom. This generated an identical stadium with two plazas, which also necessitated the removal of the Empire Pool to guarantee safe egress. In March 1996, the Millennium Commission agreed to support this development and so the ability to fund the project was available and the challenge to reconstruct the National Stadium at Cardiff Arms Park was underway, with the World Cup the movable target in 1999.

THE REDEVELOPMENT OF THE STADIUM WITHIN THE CAPITAL CITY

Cardiff Arms Park is unique in the UK and possibly in Europe as a major stadium location. It is located right in the city centre and alongside the main public transport interchange. All commercial developments are dependent upon location and access.

The site was examined to ascertain its viability related to population access within the UK and Europe. With main line rail access to London in less than two hours and inter-connection by Eurostar to Paris, road access to all the major cities in the main populated areas of the UK within three hours and Wales International Airport within 20 minutes' drive providing access to most parts of Europe within two hours, it was determined that the Cardiff location was a good commercial investment for a new third-generation stadium.

Cardiff has developed significantly over the past 20 years. In the mid-1980s, less than 15 years ago, the site of the conference, the Cardiff International Arena, was a demolition site, as was most of the South East of the City. By that time, the redevelopment of the National Stadium had been completed. Within 10 years one of the most successful retail centres in the UK had been completed, linked to major hotels and leisure facilities to the East of the City.

This leaves the National Stadium in the centre of the South West corner of the City Centre, which is in need of major regeneration and has had no significant investment within the development period of the past 20 years. The relationship between the river and the adjacent Castle and Civic Centre provides a marvellous opportunity to develop the Arms Park site as an extension of the green space within the City Centre. When related to the Cardiff Bay redevelopment, one of the most exciting of its type in Europe, the creation of the barrage provides the opportunity to link the City Centre with the bay by water. The National Stadium site with its Riverwalk and Plaza provides the ideal location for the Riverboat station from the Bay with the opportunity to go further up the river to Llandaff fields.

Cardiff is a city that has developed from a small fishing village on the estuary to the Capital City of Wales on the back of the creation of the Docks for exporting coal and steel over the past hundred years. It has been a city of two halves split by the main railway line. The dock developed to the south and the centre developed to the North to support the docks. The docks went into serious decline with the demise of coal and the centre was further developed. Now the Bay is being regenerated with major investment it is essential that this does not detract from the need for further investment in the City. The Stadium is the catalyst to maintain the balance.

The investment in the Bay is based on attracting visitors and the Stadium development has the same key requirement. It is not just a venue for a limited number of events per year, the Stadium and the facilities around will become a major, if not the major, tourist attraction in Wales. Cardiff needs to double its tourism investment and employment, needs to treble its overseas visitors to 420,000 pa, and quadruple its overseas visitors, 60% increase in UK visitor spend to £108M per annum and 25% increase in leisure day visits to 5.6M pa increasing leisure day spend by 60% to £72M per annum. To support this, significant accommodation will need to be provided and during the past two years major investment has occurred in hotels large and small, two star to five star and upgrading of existing to support the confidence generated by the Bay and Stadium developments.

Cardiff is a magnificent Capital City, the Castle right in the centre, grounds that stretch up the river for five miles, a Civic Centre accepted as one of the best in Europe.

The Stadium is not seen as operating in isolation. It must interact with the other major venues in the City. If Cardiff is to become a truly European City visitors must have the opportunity to go to a major event at the Stadium, visit the theatre, or a concert at St Davids Hall. The go-ahead for the new Millennium Centre in the Bay will be an international opera venue and the Sports village with international swimming and ice hockey facilities will add to the attractions.

The future of multi-use stadia is 'family friendly facilities', creating equal opportunities for all to attend and providing access for all.

The extension of retail is needed into the West of the City. The redevelopment of Queens Arcade is an indication that this is commencing. The retail around the Stadium is to be sports retail with specialist shopping adjacent to the transport infrastructure.

The Stadium is seen as the catalyst for regeneration for the West of the City. This commences with Central Square adjacent to the railway and bus station; £3M has recently been committed to the upgrading of the square prior to the World Cup but it is expected that the whole site will be developed after the World Cup and £10M has been committed to the upgrading of the Railway Station. None of this would have happened if the National Stadium redevelopment had not generated to invest.

Going around the regeneration area anti-clockwise, monies have just been committed to the Mill Lane Café area, creating a European Café area alongside the Marriott Hotel which has proved very successful and will be the basis of the fit-out of similar areas within the City. The areas around St Davids Hall and St Johns Church are current under public opinion review and the £6M being spent on the old library creating a new arts centre will be a major facility to help regenerate this area. Historically, the West area was a key area of the City full of Victorian arcades. Already, upgrading has occurred in these areas but the new investment will enhance this work and provide wonderful covered links between different parts of the City, and the redevelopment of the Stadium area will create the opportunity to continue this philosophy. Alongside the Stadium is a significant residential part of the conservation area of Westgate Street. The confidence engendered into the area by the Westgate Street Entrance to the Stadium, will encourage investment in those properties.

I am frequently asked 'How many car parking spaces are you providing?' I answer 'None'. The whole strategy of city centre redevelopment is to restrict vehicular access whilst improving public transport to create a quality of environment for the benefit of people's enjoyment. Twenty years ago, Queens Street to the East of the capital, had eight lanes charging down it, now it is now fully pedestrianised. The plan is to provide car parking on the periphery of the centre and park-and-ride facilities outside this. Currently, the centre has 30,000 spaces and with public transport this is expected to support the new Stadium.

This is a commercial venture, the days of public monies supporting public facilities have long since gone. The benefactors will be all operating within the city centre including the traffic wardens. Hopefully, through tourism the whole of Wales will benefit.

Communications are the key to new world of leisure. The new Stadium is seen as the centre of satellite linkages with the world as well as providing the basis for local radio and television for Cardiff and Wales.

The City Centre location of the Stadium creates a major challenge for access and emergency egress. Whilst the redevelopment has created opportunities to overcome the lack of space around the bowl the site is still restricted in comparison to a green field site or a site with space all around. The facility to create ramps, which is the ideal solution similar to the Amsterdam Arena, is not available. The redevelopment of the surrounding areas, which is critical to generate life and excitement all year round, has to ensure that the

Stadium requirements to satisfy the egress requirements of the safety certificate.

The construction of the new Stadium has generated the requirement for nine independent developments, three off site and six integrated into the site area.

Firstly, to accommodate the construction of the Stadium bowl the adjacent T.A.V.R.A., BT and DSS buildings have to be demolished and replaced at a cost of £20M. Works commenced on the provision of alternative buildings in January 1997 and have all been successfully completed.

In addition, detailed negotiation has occurred with three adjacent owners including Cardiff Athletic, County Club and Welsh Water for the reconstruction of the new pumping station which has been designed below ground with the new plaza constructed over it.

Three major developments create the exciting activity around the Stadium and adjacent areas for access.

Firstly, Millennium Square and the Scott Road Development provides an international square for the city which will develop its own identity and activities for the people of Cardiff and Wales. The development has a ten screen multiplex cinema complex as its core, creating a throughput of 300,000 people per year. This is supported by other leisure uses, with bars and restaurants, all designed by Cardiff Architects, Burgess Partnership.

Secondly: The Riverwalk, bridge to bridge, has been designed by HMA, again Cardiff architects. The city has never had the benefit of access to its river and this Riverwalk will not only service the Stadium but add to the excitement in the overall development and provide an avenue along which to access major events. The construction of the balcony section of the Riverwalk is well underway and piling has been completed adjacent to the Stadium.

Thirdly: The Westgate Development which will be created on the site of the old East Terrace and the DSS building. Both Scott Road and Westgate Street a Brunswick/Taylor Woodrow project, in this case the designers are Dobson Partnership again a Cardiff architect. The facility has at its core the World Rugby Experience and Roll of Honour which as a visitor attraction centre will generate 500,000 visitors per year which will be linked to the Tour of the Stadium and other attractions in Cardiff.

We are three years closer to our dream when we made the first submission to the Millennium Commission in 1995 with less than two years to go to the World Cup. With the enthusiasm and support to date, we will get there. See you in October 1999!

25 WEMBLEY – THE NEW NATIONAL STADIUM: DEVELOPMENTS TO DATE AND PLANS FOR THE FUTURE

B. HEAVER

The English National Stadium Trust, Wembley, UK

SUMMARY: The selection and appointment of the architect for the new National Stadium at Wembley in April 1998 will mark the visible start to this exciting project. However, in reality, a great deal of work has already been done behind the scenes in putting into place the finance, business plans, approvals, guarantees and a hundred-and-one other necessary details. This article outlines the main steps along this difficult path.

Keywords: event owners, funds, National Lottery, planning, stadium, Wembley.

INTRODUCTION

In sporting terms, it could be said that the New National Stadium is not yet in the dressing room. In fact, contrary to much media speculation, the team has not even been picked. Hence, the lists of supposed features that can be downloaded from Internet sites on this historic venue – such as the repositioning of the famous twin towers, the re-orientation of the axis through 90°, the use of the outer skin as a projection screen for laser images, the mobile transparent roof panels and the undulating 'Wembley wave' upper tier – are, at this stage, all guesswork. For, until the architect is appointed in April 1998, there are few, if any, aspects that can be regarded as fixed and immutable.

HISTORY

The present Wembley Stadium, arguably the most well-known stadium in sport anywhere in the world, dates from 1923. The first event to be held there was the 1923 English Football Association Cup Final when Bolton Wanderers defeated West Ham United 2-0, a match which reportedly was attended by 200,000 spectators. The stadium was built at a cost of £¾ million in about 300 days by Robert MacAlpine, since when it has been the venue for hundreds of national and international events. The 1948 Olympic Games were staged there, as were the 1966 World Cup, the 1995 Rugby League World Cup and the 1996 European Championships. Every year, the English Football Association Cup Final is played at Wembley, the match being televised live to at least 50 countries – more than any other league cup final anywhere. In addition, Wembley stages concerts, such as the seven

Stadia, Arenas and Grandstands, edited by P.D. Thompson, J.J.A. Tolloczko and J.N. Clarke.
Published in 1998 by E & FN Spon, 11 New Fetter Lane, London EC4P 4EE, UK.   ISBN: 0 419 24040 3

shows by Michael Jackson in 1988 which attracted a record-setting total audience of 504,000. The range of sports includes rugby league, rugby union, athletics, boxing, cricket, showjumping, speedway, hockey, American football and greyhound racing. In a more sombre role, the stadium has been used by both the evangelist Billy Graham and the Pope as head of the Roman Catholic Church.

Fig. 1. The 1924 FA Cup Final: Bolton Wanderers vs West Ham United – the 'White Horse' final.

Fig. 2. The 1948 Olympic Games.

STATUS

After 75 years, Wembley Stadium is still thriving. However, as anyone who has used the facilities will testify, there is a lot that needs improvement. For a start, the stadium's capacity, originally for a maximum of 126,500 spectators – 91,500 of whom were standing – has shrunk to 79,045 in the wake of the all-seater regulations. (During the conversion to all-seater status, the medical centre was extended, the catering facilities were extensively upgraded and expanded, and a new box office was created.) Furthermore, many of the seats have poor sight lines for football mainly because of the roof supports but partly because of the intervening greyhound track and the shallow rake of the forward seating. On the plus side, the site enjoys excellent road access and has its own London Underground station – Wembley Park, with two services (the Metropolitan and the Jubilee lines) – whose capacity could easily be expanded. One apparent obstacle to any redevelopment of the site, however, is the fact that it is Grade II Listed by English Heritage. Normally, this would be a bar to anything other than minor refurbishment but in this case English Heritage will agree to allow demolition of all but the signature twin towers and the building that links them, provided that a stadium of architectural merit is provided as a replacement.

Fig. 3. Diwali Festival of Light – October 1996.

THE NATIONAL LOTTERY

Early in 1995, Britain's Sports Council met the governing bodies of rugby league, football, and athletics to discuss the need for, and possibility of, a new national stadium. The verdict was a unanimous yes, and it was further agreed to seek National Lottery money to finance it. The proposed super-stadium would seat 80,000 and would offer unhampered sight lines for all three sports. But there was a snag: the present national stadium at Wembley was privately owned and so not eligible for Lottery funds. Four cities decided to bid – Birmingham, Bradford, Manchester and Sheffield – with London left out in the absence of a central authority such as the former Greater London Council. The latter problem was solved by setting up The English National Stadium Trust, a charity which, being at arms length from Wembley Stadium Limited, could attract funds from the Lottery and other sources. London (Wembley) accordingly joined the four other bidding cities.

After two years, the Sports Council whittled down the list from five to two – Manchester and London – and soon afterwards London emerged as the winner.

With the venue selected, the hard negotiation began and, on 6 October 1997, an initial grant of £21.5 million was issued to develop the project to the point where orders could be placed. At the same time, the grant was geared to a series of 'milestones' – some might say millstones – which had to be reached by particular dates. For example, by the end of March 1998, the Trust must:

- have negotiated a 50-year lease of the site from Wembley Stadium Limited
- have negotiated 20-year contracts with the governing bodies of football, rugby league and athletics
- be able to demonstrate support for the necessary planning permission
- be able to demonstrate that the entire scheme is soundly based, with a workable business plan, and not simply an architecturally monument.

In other words, to use the adjective that is fast being recognised as the key to all stadium and arena projects, the New National Stadium must be clearly *sustainable*.

The other principal milestones set by the Sports Council are:

- April 1998: appoint architect
- Autumn 1998: final planning consent
- Early 1999: appoint main contractor
- June 1999: begin demolition
- Early 2002: first event in the new stadium

THE NEXT STEP

As this paper is being written (January 1998), the Trust is seeking a chief executive, to be recruited for a period of five to seven years, whose job it will be not only to oversee the creation of the new stadium but also to run the business in its early years. At the same time, the Trust is looking at the architectural concept submissions from some 18 practices, including all of the top names in the UK. The plan is first to reduce the number to around a dozen and then to whittle these down to just six. At that stage, the six 'finalists' will be given a more detailed briefing and, in return, will be asked to set out the basis of their fees and the likely programmes. Interviews will follow, after which the winning architectural practice will be announced – indeed, it may be possible to use the Cardiff conference to make the announcement.

Fig. 4. Wembley Stadium under construction, 1923.

26 READING FOOTBALL CLUB: DEVELOPING WITHIN THE COMMUNITY

N. HOWE
Reading Football Club plc, Reading, UK
K. UNDERWOOD
Birse Construction Ltd, Northampton, UK

SUMMARY: In the UK a significant amount of interest is being generated by football clubs modernising and expanding facilities to accommodate expansion and commercial growth. This paper presents a success story with particular reference to how Reading Football Club is developing a non-controversial major new facility in partnership with local government and with the full support from the townspeople. Keywords: Construction, contaminated land, football stadiums, partnerships.

INTRODUCTION

Success stories of partnerships between local business, local government, sport and the community rarely make national headlines. Yet the impact on the local environment and people is major and the resulting benefits to all concerned cannot be overstated. Such success stories are even more impressive when associated with a local football club that has yet to enjoy the status of Premiership football. During the past 101 years, Reading Football Club has played and lived at its current Elm Park ground. Compared with many of today's facilities, Elm Park cannot be classified as luxurious. This Victorian stadium is now out-dated and does not comply with the recent requirements of the Taylor Report. Its location in the heart of a residential area, with poor access and parking, limited spectator capacity and facilities and no scope for expansion, means that receipts from its near-capacity (capacity 15,000) and regular loyal fans does not exceed operational costs. Increased attendance following the Club's recent success – Division Two Champions in 1993/94 to within a whisker of Premiership football in 1995 – has also placed a demand for additional capacity.

Football and business in partnership is now a well established principle. Some might say that this downgrades the game of football but it is very unlikely that any genuine football fan will complain about watching his or her favourite team (winning or losing) in the comfort of a modern stadium financed by business operations behind the scenes.

Most successful clubs reflect the commercial attributes of the Club's chairman. John Madejski, a multi-millionaire through his publishing interests, rescued the Club in 1990 from a precarious financial position. His vision and commercial success has enabled the Clubs dream to become reality in the construction of the new Madejski Stadium and Royal Berkshire Conference Centre. This £37m development has ensured that the Club's commercial future is secure under Madejski's chairmanship.

Stadia, Arenas and Grandstands, edited by P.D. Thompson, J.J.A. Tolloczko and J.N. Clarke.
Published in 1998 by E & FN Spon, 11 New Fetter Lane, London EC4P 4EE, UK.   ISBN: 0 419 24040 3

READING FOOTBALL CLUB AND THE BOROUGH COUNCIL

From the original concept, it was obvious that development of the existing Elm Park Ground was not feasible. From the start, Reading Borough Council was supportive of the Club's plans with emphasis being placed on a new development that would benefit the community as a whole.

The Council, favouring a location for the new stadium away from any residential areas but still within the Borough boundary, made available a landfill site at Smallmead on the outskirts of Reading. The site, known to Reading residents as the local rubbish tip, was strategically placed for access but in itself raised technical and commercial challenges of site remediation before use (Fig. 1). The 66-acre site, half a mile from Junction 11 on the M4 motorway and only two miles from the town centre, also offered the opportunity to raise finance by selling part of it (16 acres), after remediation, to Salmon Harvester for a 20,000 m² non-food retail development. Funds received from the sale contributed a considerable proportion of the costs of remedial work and stadium construction costs.

Co-operation with the local Council resulted in Reading Football Club's receiving overall planning permission with safeguards to stop retail development going ahead without the stadium. For Reading Football Club, the cost of land remediation meant that the retail

Fig. 1. Land reclamation nearing completion.
(Photo reproduced by permission of Cloud 9 Photography, Reading)

development was a vital and integral element of the financial plan. The new development will also benefit the local community in a number of other ways:

- the existing Elm Park Ground has been sold for much-needed social housing
- the new development has regenerated a contaminated site
- the development car parks, second only in size to Wembley, will provide space for over 2,000 cars. Part of this area (on non-match days) will provide the community with valuable park-and-ride facilities which will operate in reverse on matchdays
- the location of the new stadium and associated new relief road directly to Junction 11 on the M4 will alleviate traffic problems on match days along the A4 between the M4 Junction 12 and the existing Elm Park Ground
- the ground and stadium at the new site will be available for community events
- the scheme also plays a major role in the long-term plans to enhance the whole of the south-west Reading area
- three sports pitches will be used for team taining and community teams at other times
- creation of further employment
- supply of essential conference facilities for the rapidly growing number of local businesses.

THE MADEJSKI STADIUM

The final go-ahead for the Madejski Stadium was given on 29 April 1997, triggering construction of the 25,000-seater stadium. Construction began in June 1997 and is on schedule to give Reading Football Club state-of-art facilities to rival those at any club in the country in time for the 1998–99 season. The basic design concept required:

- a visible and stunning landmark
- incorporating the latest technology and safety features
- a state-of-the-art design with fully enclosed seating bowl
- suspended roof for optimum spectator vision
- dynamic building form
- focus on spectator comfort and facilities
- operation seven days a week
- comprehensive hospitality accommodation
- modern information system.

The resulting design presents spectators with optimum viewing facilities, with high-quality seating, a fully enclosed structure and suspended roof, with floodlights along the front canopy. The stadium comprises single tier north, south and east stands with a ground-level concourse housing varying facilities. Within the two-tier main stand, excellent corporate facilities are provided with 28 executive boxes, two corner lounges and a luxurious 'view-of-the-pitch' restaurant. The stadium concourse will house 12 bars and a variety of hot- and cold-food retail outlets. The main stand will include a spacious superstore where supporters can purchase a wide range of club merchandise complemented with smaller outlets in each stand where mail order items can be ordered and match day programmes purchased.

Fig. 2. The new stadium and conference development.

There will also be a junior clubroom, where children aged two and upwards can stay, fully supervised, while their parents enjoy the match.

Altogether, these facilities will provide spectators with an infrastructure to support their enjoyment and comfort during their visit to the stadium. The club benefits from a satisfied customer base and a direct financial income – so important to its ongoing commercial success.

Construction contract summary

Work began on site on 2 June 1997 and the stadium has to be completed to enable Reading Football Club to re-locate from its existing ground on 31 August 1998.

The remediation of the site is being carried out under ICE 6 Contract with the infrastructure and stadium works under a JCT 1981 Design and Build Contract. An 'umbrella' Guaranteed Maximum Price Agreement of £25.5 million sits over the three separate contracts: reclamation works £5.3 million; infrastructure £2.2 million: stadium £18.0 million.

Outline design requirements

The stadium will conform to the recommendations of the Taylor Report, the Guide to Safety at Sports Grounds (Green Guide), and the Football Stadia Development Committee (FSDC). In addition, it will incorporate the latest recommendations on spectator safety and comfort. All aspects are designed in accordance with appropriate national standards.

Site preparation

The main objective of the design strategy was to facilitate the redevelopment of the new site without allowing the environmental conditions on or around the site to deteriorate as a result of any activity.

The stadium is being constructed on a former landfill site used largely for the disposal of domestic refuse between 1975 and 1985. Before development, the ground conditions consisted of a clay capping overlying 4–8 m of domestic refuse below which were mixed bands of sands, gravels and clay.

Site remediation and re-profiling was carried out to a design by Ove Arup & Partners as part of an earthworks contract that involved installation of leachate drains, methane venting and a clay capping beneath the stadium site.

Fig. 3. The stadium under construction in November 1997.
(Photo reproduced by permission of Cloud 9 Photography, Reading)

The project involved the excavation of 490,000m^3 of waste and associated materials from the north-eastern part of the site. This area covers approximately one-third of the total development area. The excavated waste was spread and compacted over the south-western two-thirds of the site. This waste was placed to form a plateau that formed the level for the stadium car park. The result of this activity was that the northern part of the site became a clean uncontaminated void for development. This void was then backfilled with clean engineering material before the construction of the retail park.

The stadium area takes the form of a raised concourse combined with a low-level pitch construction. This approach reduced the amount of waste movement and corresponding height gains. The height factor was also an important consideration from a visual intrusion aspect. The overall plateau has been raised some 6–8 m and has required careful design and construction of reinforced earth retaining walls around the plateau perimeter.

Owing to the nature of the site, and in particular the potential for considerable ongoing ground settlement, the superstructure to the stadium, the ground-floor slab and the below-slab drainage are supported on piled foundations.

Substructure: ground floor construction
The ground floor construction incorporates a methane barrier and passive venting system

to counter any risk of methane passing through the primary protection system within the site capping below. In developing substructure design proposals, efforts were made to minimise the extent and depth of excavation into the clay capping beneath the stadium. The floor construction comprises a 275 or 300 mm power-floated in-situ reinforced concrete slab on an Aldaprufe GRA high-performance methane barrier membrane on 200 mm Cordek Ventform units to create a void beneath the slab. The slab spans 7–7.5 m between main ground beams.

Cross ventilation is provided to the below slab void by two periscope vents through the edge beam to the rear of the stand and through a terrace riser at the front in each bay along the stands.

Concrete for the ground beams, pilecaps and floor slab will be grade C35 with mix proportions suitable to provide resistance to Class II sulfate levels.

Superstructure: floors, staircases, terrace units
Upper floors within the stand are of 130 mm-deep mesh-reinforced concrete slabs cast on profiled metal decking spanning unpropped between secondary beams at 3–3.5 m centres. The supporting steelwork is designed to act compositely with the floor slabs by means of through-deck welded shear studs.

Stair flights throughout are precast with galvanised tubular steel handrails designed for crowd loading as appropriate.

Terrace units are precast concrete cast in steel moulds. The treads are at least 100 mm thick to accommodate cast-in barrier fixings and also to incorporate a 20 mm crossfall for drainage in accordance with FSADC Guidelines.

Roof structure
A prismatic steel truss acts as the primary support system to the roof, which is supported at 12 positions around the stadium on masted frame structures.

The concept of all the masted frame structures is similar, being that of a back-stayed cantilever, with tubular steel ties and struts providing support to the primary truss. However, their visual form varies according to position, with those at the four corners of the stadium being significantly larger than the other eight to give greater visual impact. The roof is clad with profiled metal sheet decking supported on cold-rolled steel purlins which in turn are supported on cellular beam rafters. The latter are connected to the underside of the prismatic truss. A system of bracing in the plane of the roof transfers stability and in-plane wind forces back to the supporting structure.

CONCLUSION

This paper has shown how a 'rubbish tip', and what was once perceived as a liability, is rapidly being transformed into a public and commercial success. The story of the development of the stadium with associated conference facilities is not yet complete. At the time of writing, a planning application has been submitted by Reading Football Club for the further development of the site by the construction and operation of a major hotel facility integrated with the new exhibition and stadium complex.

27 DEVELOPMENT AND MANAGEMENT OF FACILITIES AT THE VALLEY STADIUM – BUILDING SUCCESS

I. THIRLWALL
James Consulting Engineers, London, UK
R. KING
Charlton Athletic Football Club, London, UK

SUMMARY: This paper provides an overview on The Valley Stadium's role at the centre of Charlton Athletic Plc's success as a sport and leisure business in the community. It describes its humble beginnings in 1921 and major events such as the 1947 F A Cup Win; the dark days of the ground closure and ground sharing, then the return to The Valley and its 'renaissance' made possible by investment and vision from the fans and Football Trust grants. The new East Stand, followed by the phased development of the West Stand, together with the development of the sport and leisure business and the partnership with the local council and community for the regeneration of Greenwich and Thames Gateway through the Millennium exhibition, in creating employment to manage all the facilities at the stadium, and continue its development as a Stadia for 2000 and beyond.
Keywords: Football stadia, management, partnership, redevelopment.

Figure 1. East Terrace: Charlton -v- Millwall 1927.

Stadia, Arenas and Grandstands, edited by P.D. Thompson, J.J.A. Tolloczko and J.N. Clarke.
Published in 1998 by E & FN Spon, 11 New Fetter Lane, London EC4P 4EE, UK.   ISBN: 0 419 24040 3

INTRODUCTION AND INITIAL DEVELOPMENT

The Club was founded in 1905 by a group of local lads who played their first game in the Woolwich and District League. They joined the Football League in 1921 and played their first League game at The Valley on 21 August 1921 against Exeter City AFC.

The Stadium started to take shape in 1924 when the first West Stand was built; it provided seats for the directors and 500 spectators. The North Stand was constructed in 1935 and became the favourite end for the Addicks Fans. The South Stand for away team supporters was erected in 1979. The only uncovered spectator area of the ground was the East terrace which provided space for 45,000 standing spectators, in the days when the Club was almost invincible with players like Sam Bartram, Eddie Firmani, Stuart Leary, Don Welsh, Chris Duffy and Derek Ufton under the shrewd management of Jimmy Seed.

In 1947 Charlton won the F A Cup and during the 1940s and 50s they were a side of strength and character. In this period they moved from the 3rd to 1st Division in consecutive seasons winning the 3rd and 2nd and being runners-up in the 1st Division – an unequalled record! Like so many other clubs of the time, little or no money was spent on the stadium for either the safety or the comfort of the spectators and the players were no better off with their £20 a week maximum wage. Those were the days! No Match of the Day, no club shops, no home and away strips and everybody was asking what happened to the gate receipts?

THE WILDERNESS YEARS AND RETURN TO THE VALLEY

The sixties and eighties were not the best of times for Charlton and culminated in them having to leave the Valley in 1985 for a variety of reasons that included poor financial management and unsafe conditions for the spectators on the East Terrace.

As a result, Charlton were one of the first Clubs to ground share – with Crystal Palace at Selhurst Park and West Ham United at Upton Park.

Throughout this time of nomadic existence their faithful fans never gave up hope of returning to the Valley. Amongst them were a group of young directors who formed the new board of directors and invested sufficient money to enable the Club to return to the Valley in 1990 to start the clean up operation and erection of temporary stands, offices and hospitality areas. After 18 months of blood sweat and tears, Charlton Athletic returned to the Valley to play a football league game on 5 December 1992.

THE TAYLOR REPORT AND A RISING IN THE EAST

The tragic events at Heysal, Hillsborough and Bradford had preceded this occasion and the Taylor Report placed stringent requirements for the safety of the spectators on the Club who did not flinch from their responsibilities. Indeed, no sooner had they arrived back at the Valley then they unveiled their plans for a new stand on the site of the obsolete East Terrace. In March 1994 the plans were realised when a new 6,000 seat East Stand was opened.

Fig. 2. In the Wilderness.

Fig. 3. Return to the Valley 1992.

Fig. 4. The East Stand 1994.

Fig. 5. The West Stand – Stage 1, 1997.

THE WEST STAND DEVELOPMENT

A year later a temporary West Stand was replaced with what became the first stage of the double tier 8,000 seat stand; the second stage (lower tier) was completed in September 1997 along with accommodation for the new changing room, reception, Floyd's Sports Bar at ground floor level and Directors, Vice Presidents and Millennium Lounges at first floor, all beneath and behind the terrace seating area. The portacabin village of offices, changing rooms and hospitality areas were surplus to requirements just five years after they were installed as part of the back to the Valley works.

This stage of the West Stand development was carefully planned to suit the requirements of the Club and the community; with weekday and weekend use of the facilities all year round. The Club has developed the Stadium so that it can grow as a business and become the centre of excellence for leisure activities in the community. The headquarters of Greenwich Sporting Club, which is operated by Greenwich Leisure, will be within the West Stand when the upper tier is completed in September 1998. Greenwich Leisure will take over their new offices, aerobics studio, cardiovascular gymnasium, and lecture/seminar suites at the first floor level of the stand.

THE STADIUM AND THE GREENWICH PARTNERSHIP

Since their return to the Valley, the Club has taken football into the community and have strengthened their relationship with Greenwich Council's leisure department. As their business has grown both on and off the field, so they have created jobs for people in the community – the development of the stadium through careful planning and well-managed investment has lead to the planned organic growth of the business. This development will continue with the redevelopment of the North Stand.

The Valley Stadium is the headquarters of a successful, dynamic business whose core activity is football – like any other football business, its success is almost wholly dependent on the footballing performance of the team, primarily in the League, with bonuses available from successful runs in the League and F A Cups. The success is largely due to the quality of the management, the players, and the facilities of the Valley Stadium and of course the faithful fans

MANAGEMENT OF THE FACILITIES, MILLENNIUM AND REGENERATION

The facilities provide a venue and focal point from which the Club is playing its part in the Regeneration of the area – Greenwich and Thames Gateway.

The is being achieved by a working partnership between the Club and the Council (who represent the local community). The initial tangible benefit of the partnership is the creation of jobs in the area, both directly at the Club, and indirectly for suppliers of goods and services in the area (clothing, printing, food and drink etc.).

The advent of the Millennium Exhibition will accelerate the creation of jobs and be a major force in the regeneration process – investment in the transportation infrastructure, extending the Jubilee Line and Docklands Light Railway as well as improvements to the Woolwich Road will ensure that visitors and indigenous local inhabitants who live and work in the area will be able to travel efficiently and comfortably to all the 'leisure' sites (Exhibition and Heritage) at the Millennium and beyond to the London Olympics in 2008!

Greenwich Leisure HQ will be at the Valley Stadium in September 1998; the Sporting Club Greenwich is their creation and would bring together the major and minor sporting activities within the area – Blackheath RFC, Blackheath Golf Club, Cambridge Harriers etc – and enable Greenwich College to offer practical hands-on experience for the students on sports science and management courses.

The management of the facilities at the Valley Stadium are the sole responsibility of the Director of Operations and include the continuous stadium development, commercial premises, training ground at Eltham, planned and emergency maintenance and repairs to all facilities, cleaning, safety, security, match day operations and pitch maintenance. The Facilities Management 'Team' consists of the ground staff for the general building maintenance and specialist contractors for the building M & E services maintenance. Over and above these duties, the safety and comfort of the fans has to be ensured – by close liaison with the Safety Committee and the operation of a customer relations office which responds to the constant flow of the fans' verbal and written suggestions and criticisms!

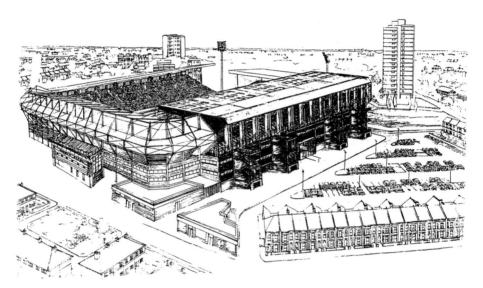

Fig. 6. The Valley Stadium in 2000.

28 MULTI-PURPOSE INDOOR ARENAS: THE NYNEX ARENA, MANCHESTER – A DEVELOPMENT CASE STUDY

P. L. JORGENSEN
Ellerbe Becket International, Kansas City, Missouri, USA

SUMMARY: The paper describes the characteristics of an indoor arena, advantages of such a facility for a city, and shows the development process for the Nynex arena in Manchester, England.

INTRODUCTION

A multipurpose indoor arena is a public assembly building designed to accommodate large crowds of people viewing a wide variety of events in a controlled indoor environment. Americans use the term 'arena' to denote this project type. We have designed forty four arenas and have learned a great deal about this building type and their interaction with cities. Events can be most anything including circus, concerts, ice shows, exhibitions, sporting events, rallies, political gatherings, traveling productions, athletics, and equestrian events. We have designed buildings from 6,000 to 25,000 seats with event schedules varying from about 100 to over 500 events per year with annual attendance ranges of a few hundred thousand to well over four million visitors. We are currently involved in projects in Europe, Russia, Asia Pacific, South America, and North America.

An arena benefits a city by adding life and economic impact if it is appropriately located and designed for the local market. Imagine an average of 7,000 people attending only 150 events in a city annually. This is over one million people a year spending money, moving through the city, and engaged in the local economy. Imagine attracting artists and acts to your city that would otherwise pass it by. Imagine a focus for local talent and culture to promote regional events that otherwise might languish. Think of the arena as an entertainment center that works within the fabric of the city. This is a place to take the family for great social experiences. It is not just a sports center.

We believe that these projects work best in the city center with good access to public transport and other city services. Central sites are not without the problems of land availability as you will need from 3 to 5 hectares for just the arena footprint. Planning permission may be difficult in a building that can be as tall as 25 meters with a volume of between 2 to 3.5 square meters per seat depending upon the configuration. Costs for the project vary considerably due to location and design brief. However, we would normally begin budgeting around $US1,500 per square meter and adjust for scope, location and complexity.

Stadia, Arenas and Grandstands, edited by P.D. Thompson, J.J.A. Tolloczko and J.N. Clarke.
Published in 1998 by E & FN Spon, 11 New Fetter Lane, London EC4P 4EE, UK.   ISBN: 0 419 24040 3

A successful facility is appropriate to its market place. It is the right size, neither too large or too small. Design decisions must be based on market analysis and operating budgets. The facility must work for the local culture and present an image that is attractive to those who will use it. Flexibility must be designed in to maximize the number of events which can be accommodated. Long term adaptability is important to facilitate alterations to the building as the market grows and changes. The arena must be located in an area that is an entertainment destination in the city. It must be properly managed, promoted, and maintained.

Successful arenas are designed and run like a business. Revenues are maximized and operations are run as efficiently as possible. The quality of the experience for the visitor is a primary concern. Revenues are generated from suites (also called boxes or loges), premium or club seats, restaurants, food and beverage outlets, merchandising outlets, advertising, naming rights, special clubs, and many other ways in addition to tickets. These revenues are extremely important. All of these features must be tuned to the local market and designed in to the building.

An arena is a natural ingredient to urban regeneration schemes. There are good sites coming available in many older cities with the privatization of rail and rationalization of other services. These locations normally have good infrastructure and excellent development potential. This makes for good partnerships between the city and the private development sector.

There are three basic approaches to building an arena project: privately financed, public-private partnerships, and public development. Successful examples of each of these approaches exist.

Few privately developed arenas are completely private sector projects. It is very difficult to support the capital costs from the operating profits except in very special markets. In most cases where the buildings are privately financed there is usually some form of public contribution in land or infrastructure. Today it is possible to find private money for these projects in many places around the world, but only under the right conditions. Where private money is involved building features and management are the key to producing believable contractually obligated revenue streams for lenders. Some of the arenas and stadia that are currently in development in the US represent highly developed methods of capturing revenue, but this is a very special market response that may not be appropriate in other places. Examples of privately funded arenas are Madison Square Garden in New York and The Palace at Auburn Hills.

Public-private partnerships have been accomplished in every conceivable combination. An arena can be the basis of a planning gain for a development project for the private sector. The best projects of this type demonstrate real cooperation and shared goals between developers and cities. You must be realistic about how much the private sector can carry and what the public purse will support. Examples of this type include the Centro Arena at Oberhausen, Germany and the Nynex Arena at Manchester, UK which will be discussed in more detail.

Public projects are easiest to understand, but no easier to implement. These too must be market-based and have a business-like approach to management, operations, and maintenance. Without this they will flounder in the market and attract far fewer people to events. Perhaps the most successful public arena in the US is the Mark of the Quad in Moline, Illinois.

MANCHESTER NYNEX ARENA

Victoria Station, Manchester has been an important rail station since 1844. It was one of three main line stations serving the city during the busiest years of the industrial revolution. The main buildings are listed historical buildings. The station occupied some 14 acres in the oldest part of the city just north of the city center. The station has lost importance to Piccadilly station and no longer required the 17 platforms and land needed in earlier days.

The station location is ideal for both public and private transport. The station serves as a major bus station, taxi stop, a tram stop, and a main line station. It is adjacent to the ring road around the city centre. It would be difficult to find a more accessible site in Manchester.

Intercity Property Group (IPG), a private development company, first looked at development opportunities at the station in 1988. An arena was a part of the earliest schemes. IPG worked with British Rail (BR) and Manchester City Council to determine their needs as part of the development strategy. The site fit the urban regeneration goals of the city and BR was interested in disposing of excess land at market rates. A proposal was developed that included 42,000 square meters of office space in a 180 m tower, a five star hotel, the arena, a completely rebuilt station, a new station concourse open to the public with retail, commercial, leisure uses, and a multistory car park.

The concept was received enthusiastically by both the City and BR. These discussions cemented a partnership among all of the groups which was to prove essential and exists to this day. However, it became clear that with the increasingly depressed property market of the day and BR's financial and operational constraints prior to privatization that to bring the scheme to fruition required public sector support. The level of support needed indicated Central Government support was necessary. As a result the scheme languished for some time.

In 1991 the Central Government was focused on the Manchester 2000 Olympic bid. In August of that year representatives of the City Council and Manchester 2000 asked IPG to put their scheme on the table again with an eye toward moving it forward quickly with a larger arena to accommodate 16,500 seats for gymnastics finals. Central Government had pledged a total of £55 million to the Olympic bid and would provide an appropriate amount of this funding for the arena.

It was necessary for BR to market its surplus land at the station. This open competition required IPG to bid for its own scheme in January 1992. The competition was fierce, but IPG prevailed and was designated the preferred developer (Vector Investments Ltd. which included Bovis Construction) in June of 1992. A planning application was made in August 1992 and detailed consent was granted with conditions on 29 October 1992. On 4 December agreements were signed to lease the land and air rights for 199 years and to receive the largest UK Government City Grant given to that time of £35.5 million including £4 million for reconstruction of the station. Manchester City Council agreed to provide £1.5 million for transport aspects of the scheme. Bovis Construction was appointed management contractors and agreed to a phase one guaranteed maximum price. Preliminary work started on site that day. The Olympic Technical Committee arrived on site in March 1993 to see a major commitment to the 2000 bid well under way. The concept of the arena was praised by the committee.

The phase one project included:

- Arena: 16,500 ice hockey; 20,000 center stage concert (the largest in Europe), open July 1995
- Multistory car park: 1,050 spaces, 50 reserved for BR needs
- Multiplex cinema: 10 screens, 1,600 seats
- Arena Point: office block of 2,600 square metres of commercial lease space.
- The City Room: 2,300 square metres of retail commercial space (expandable to 9,200 square metres)
- The station: four through lines (previously seven) to the most modern standards, open April 1994

The first phase is open and leased. Private funds were provided by the Cooperative Bank whose corporate headquarters overlook the site. There were bids received from private management companies for the operation of the arena. Ogden Entertainment Services, a division of a US Fortune 100 company, was selected. Though Ogden manages more major assembly facilities than any other group world wide this was their first building in Europe. They have signed a 20 year agreement offering Vector an index linked annual payment and a share of gate receipts. Manchester City Council have agreed to manage the car park for 25 years offering an index linked annual payment.

The arena from 15 July though year end in 1995 hosted 66 major events which drew 535,454 people. This is remarkable for a new building in a market which previously had no comparable facility. In 1996 the facility held 135 events and drew 1,015,211 people. 1997 should produce 183 events and even larger attendance numbers. None of these numbers includes attendance at non-ticketed events such as exhibitions or rallies where the facility is leased for a fee.

What was the city really looking for in the scheme? Positive action on regeneration goals, use of and development of an existing transport node, and a gateway (tower) marker on the undeveloped north side of the city center. The developer was looking for a developer's profit. BR was looking for market value for excess land. Central government was looking for a quality quick response on the Olympic bid, something tangible other than marketing. Everyone worked together in partnership to get what they wanted. There was a strong sense of team work and commitment to the project by all parties.

Total development cost for phase one was £72 million. The grant was actually a deficit grant based on an open book appraisal of the scheme including a reasonable developer's profit based on commercial rents. The grant filled the deficit.

What impact has the facility had? There are no hard scientific numbers and a University study is under way at the moment. However, there is clear evidence of rising land values around the arena and the catering businesses in the vicinity have done extremely well. In an area that was derelict, there is now a rising market. After an event you see many people dispersing into the city for dinner and drinks. There is more street life at night in a city that was already well known for it. Attendance should continue to grow over time as the building develops its reputation as a place for entertainment to stop on European tours and as indoor sports continue to gain in popularity. The local use of the facility for Manchester events will continue to expand. However, it is clear that the building is already established in the minds of Mancunians as a part of the local culture.

How can other cities learn from this experience? The city must have a vision for what kind of city they want to be. They must have the determination to achieve their goals. They must have leadership in both the public and private sectors that produces the self confidence necessary to accomplish big things. There must be a dialogue to find common objectives that everyone can support. Everyone must work for the common good, not just their own self interest. It requires partnership.

The design team for the project included:

* Austin Smith Lord – master plan architects
* DLA Ellerbe Becket – arena and commercial architect
* Ove Arup & Partners – engineers
* DLE – Quantity Surveyors

This teamwork which was so important to the arena project has been called on again with the IRA bombing of the city center in June of 1996, less than a year after the arena opened. It is the same public and private leadership which has responded to this crisis. Though it devastated the city center, there was a new plan developed by competition and approved less than six months after the bombing. A year later and construction is well underway on a redeveloped city center which will strengthen the links to the arena.

SUMMARY

The key points in a successful arena development:

* Committed public and private sector leadership
* Good site – right location and size (3–5 hectares)
* Market analysis and business and operations plan
* Experienced team of experts in development, design, construction, local issues, building operations
* Realistic development and financial plan.

29 THE REDEVELOPMENT OF TWICKENHAM

D. A. WEBSTER
Mott MacDonald, Croydon, UK

SUMMARY: The redevelopment of Twickenham, which commenced on site in May 1989 and was completed last year, has totally transformed the ground into a major sporting venue with an all-seated capacity in excess of 75,000 spectators. A new three-tier stand with a 39 m cantilever roof has been constructed in three phases around the north, east and west sides of the ground. This paper describes the major design features and a number of construction techniques adopted and includes the background to the development, together with several peripheral issues that need to be addressed in stadia design.

Keywords: Cantilever, connections, construction, corrosion protection, costs, design, dynamic testing, environmental effects, fire engineering, foundations, roof, sheeting, steelwork, wind tunnel testing,

INTRODUCTION

In 1988 the Rugby Football Union decided to redevelop Twickenham as a major sporting venue, having previously elected not to relocate but to rebuild at the home of rugby football. The plan involved replacing the North, East and West Stands in a phased manner and linking up to the rebuilt South Stand which had been completed six years earlier (Fig. 1).

The fundamental requirement was to provide the maximum amount of spectator accommodation within the practical constraints of the site. In the case of the North Stand, which formed the first phase of the redevelopment, the proposed capacity was for 20 000 spectators in a combination of lower standing and upper seated terraces. This requirement was achieved by adopting a three-tier stand solution which, when continued around the ground along the east and west sides, would have produced a final total capacity of 100 000 spectators.

The tragedy at Hillsborough in April 1989 changed the original concept of providing both standing and seated areas when, within a few weeks of the event, the RFU elected to change to all-seated accommodation anticipating Lord Justice Taylor's Report by some 12 months. The resulting final total ground capacity is now 75,000 spectators which includes approximately 15,000 seats in the North Stand, 25,000 seats in each of the East and West Stands, and 10,000 seats in the South Stand.

Stadia, Arenas and Grandstands, edited by P.D. Thompson, J.J.A. Tolloczko and J.N. Clarke.
Published in 1998 by E & FN Spon, 11 New Fetter Lane, London EC4P 4EE, UK.   ISBN: 0 419 24040 3

Fig. 1. Aerial view of the original ground.

The final requirements of the design brief included:

- maximum number of seated spectators
- optimum sightlines and visibility
- unobstructed views
- appropriate protection from the weather
- a good standard of seating and range of facilities for both able and disabled spectators
- the maximum amount of hospitality boxes
- the maximum volume of space within the stand for other accommodation
- a low maintenance regime
- the maximum possible ground capacity at all times during redevelopment
- the minimal adverse effects on the condition of the pitch.

It was recognised that a number of these requirements related to each other and that the design would need to be developed to ensure that the final solution, as far as practical, fulfilled all these objectives.

DESIGN SOLUTION

The three-tier arrangement shown in Fig. 2 was selected to maximise spectator capacity whilst providing optimum visibility by minimising sightline distances. All spectators are contained within a 185 m radius from the corners of the playing area. The highest seat was

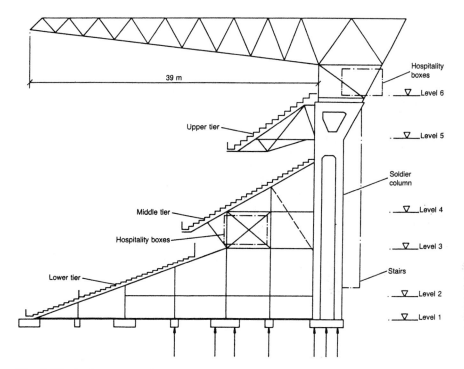

Fig. 2. Typical cross-section showing concourse levels and structural arrangement.

limited to 30 m above ground level and escalators and lifts are provided to gain comfortable access to the upper levels in addition to the stairs located at the rear of the stand feeding into concourses. In particular, major expanses of broad stairs to the rear of all stands were conceived to provide ready access to the large level 2 concourses and hospitality facilities.

Fully serviced concourses are provided at all main access levels to the three tiers of seating and run continuously for the full length of the new stands.

Areas for disabled spectators are provided at the front of the lower tiers accessed directly from ground level. Disabled facilities accessed by lifts are also located at the rear of the lower tiers in the southeast and southwest corners of the stadium where the new stands are wrapped around to meet the existing South Stand. Hospitality boxes are generally provided for the full length of the new stand at the rear of both the lower and uppers tiers. Associated facilities are located in the lower levels of the stands taking full advantage of the large volume of space under the middle and lower tiers.

A cantilever roof was selected to provide unobstructed views and to facilitate a phased development by extending the North Stand roof, constructed as the first phase, along the east and west sides of the ground in subsequent phases (Fig.3). Although other alternatives were considered, it was evident that the cantilever solution was particularly appropriate for maintaining a uniform seating arrangement around the corners of the ground and thereby maximising the overall ground capacity.

Frames to support the three-tier arrangement and the roof are provided at 7 m centres and include concrete 'soldier' columns at the rear of the stand which form a dominant feature of the rear elevation.

Fig. 3. Completed North Stand and part-completed East Stand redevelopment.

ROOF CANTILEVER

Steelwork

A steel cantilever truss roof spanning 39 m was selected to provide an appropriate degree of spectator weather protection. Whilst a greater cantilever span would clearly have provided an increase in spectator weather protection, a span of 39 m was considered to provide the optimum protection when taking into account the construction cost and the shadow effect of the roof on the condition of the pitch.

An alternative roof design solution utilising a longitudinal girder arrangement was considered, but would have presented practical difficulties with the wrap-around design concept. Cable-stay solutions were also investigated but were not considered appropriate for the size of roof span required, particularly since wind uplift on the roof is a critical load case and the concept of cable-stays in compression is not an ideal structural arrangement.

The roof steelwork generally incorporates grade 50 circular hollow sections and consists of plane trusses at 7 m centres, tapering in depth from 6 m at the rear to 1.5 m at the leading edge (Fig. 4).

Longitudinal restraint to the roof is provided by plan bracing in at least two bays per stand, together with the strut and tie action arising from the wrap -around shape. Horizontal forces are transferred via longitudinal members to the central bays of each stand where vertical bracing takes the forces down to ground. Movement joints are provided through the full 30 m height of the structure at the junctions of each stand.

The tips of the cantilevers are jointed by a 1.5 m-deep inclined stiffening fascia girder designed to minimise the potential differential wind effects at the free edge. The trusses are

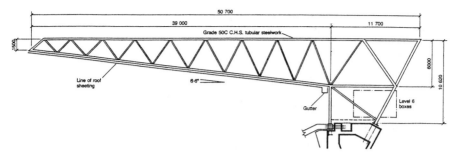

Fig. 4. Typical roof truss.

jointed together at the rear of the stand by a triangular space girder to form the pyramid roof shape which dominates the appearance of the stand, particularly on the rear elevation (Fig. 5). The purpose of the girder is both aesthetic and structural, since it contributes to the robustness of the structure and assists in damping possible vibration of the roof trusses arising from any potential adverse dynamic wind effects.

Sheeting

The roof sheeting is located at the underside of the trusses both for aesthetic reasons and to comply with planning requirements to reduce the perceived height of the stadium. The roof falls at 6.6° and is drained to the rear of the stand, where water is collected in a substantial gutter and transferred to ground in downpipes cast into the rear concrete soldier columns.

Fig. 5. Rear elevation.

The roof sheeting is general Filon translucent GRP to provide natural light, and is supported on grade 43 rectangular hollow section purlins at 2.3m centres spanning beneath the lower booms of the trusses (Fig. 6). Atlas profile plastic-coated steel sheeting is used in 1.8 m wide strips at each truss to provide walkways for inspections purposes.

Fig. 6. Underside of the roof.

Roof purlin spaces are reduced to 1.7 m towards the front edge of the roof to suit increased local wind effects. Rectangular hollow section steel purlins were selected to minimise maintenance and the possibility of pigeons perching on the underside of the roof, an aspect which is known to be a significant maintenance problem in many stands.

UPPER, MIDDLE AND LOWER TIERS

Upper tier cantilever
The upper tier of terracing consists of precast concrete units supported generally on grade 43 steelwork cantilevered frames at 7 m centres which are bolted, using 32 and 25 mm diameter Macalloy stainless steel prestressing bars, into the main rear concrete soldier columns. The cantilevered span of the terracing is approximately 12 m and the design of the support frames included an assessment of the vibrational response in relation to the likely dynamic loading caused by spectators. The tier was designed to have a loaded vertical natural frequency in excess of 7 Hz and in addition precautionary measures were adopted to assist in damping the induced vibrations. These included the introduction of shear connectors between the floor at level 5 and the support frames to mobilise the stiffness of the concrete floor slab. It is now a requirement of the recently published Fourth

Edition of the *Guide to Safety at Sports Grounds* [1] ('The Green Guide') for dynamic assessments of structures to be undertaken where seating decks have a vertical frequency of less than 6Hz or a sway frequency of less than 3 Hz.

Middle tier

The middle tier of terracing also consists of precast concrete units supported on grade 43 steelwork support frames at 7 m centres constructed above the lower in-situ concrete frames. The lower and middle tier frames provide the necessary lateral support of the rear concrete soldier columns, which in turn support the cantilever roof and cantilever upper tier. As for the roof and upper tier, stainless steel Macalloy prestressing bars are used to connect the middle tier steelwork to the in-situ concrete.

Lower tier

The lower tier is constructed in in-situ concrete. This material was chosen to overcome lead-in times for the construction of the stands and permitted opening of the lower tiers to spectators within six months of demolition of the old stands.

In-situ concrete has the natural advantage over precast in providing a more easily achieved watertight construction which is particularly beneficial in the more exposed lower tier. The choice of in-situ concrete is also consistent with the fire protection requirements where the lower sections of the stands are allocated for multipurpose usage.

STEELWORK TO CONCRETE CONNECTIONS

The roof, upper tier and middle tier main steelwork members are connected to the rear concrete soldier columns using 32 and 25 mm diameter Macalloy stainless steel prestressing bars. Typically, all such bars are wrapped in debonding tape and pass through the concrete in 75 mm diameter steel tubes utilising the I shape of the soldier columns as illustrated in Fig. 7. This arrangement provided the necessary construction tolerances required by the contractor and permitted stressing of the bars following grouting of the tubes and endplates. Full-scale grouting trials were undertaken by the contractor to ensure that the techniques and details were satisfactory.

Achieving accurate vertical alignment of roof truss tips can present practical difficulties. An adjustment facility was therefore incorporated in the design of the cantilever connection to the concrete soldier which was successfully adopted by the contractor.

The tips of the cantilevers were initially set high and then gradually lowered by releasing the nuts to the Macalloy bars at the rear cantilever/solider column connections whilst pivoting the trusses on the front cantilever/solider column connections (Fig. 4). The fascia girder was aligned after the truss tips had been correctly adjusted, whereupon the fixing bolts were then fully tightened.

FOUNDATIONS

The structure is supported on 600 and 460 mm-diameter augered bored piles, approximately 20 and 26 m long, founded in London clay. The ground conditions encountered were generally as predicted by the site investigations: 1m of variable fill material above

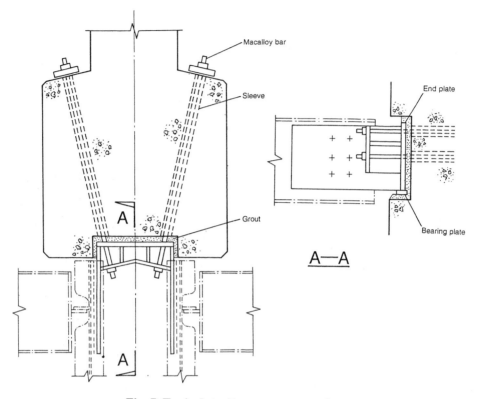

Fig. 7. Typical steel/concrete connection.

4 m of sand and gravel overlying London clay. The piles were subjected to integrity testing, and static load testing confirmed a safe working load of 110 tonnes. Pile foundations were considered to provide the most economical solution in view of the significantly high imposed loads, although pad foundations were provided towards the front of the lower terrace structure where the loadings are relatively low. The calculated differential settlement values between piled and non-piled foundations were incorporated into the design of the structural frames.

WIND TUNNEL AND DYNAMIC TESTS

As recommended by *Appraisal of sports grounds* [2] full consideration was given to the appraisal of dynamic effects on the roof and wind tunnel testing.

Natural frequency and damping ratio parameters were determined for the roof and upper tier and were used to derive a 'mildly dynamic' classification for the roof as defined by Cook [3] and Building Research Establishment (BRE) Digest 346 [4]. A 'dynamic magnification factor' of 1.25 was evaluated for enhancement of static wind loads to determine design wind forces.

Wind tunnel tests were undertaken at the BRE laboratories, Garston, Watford,

England, on a 1:250 scale model of the partially and fully redeveloped stadium to determine static wind pressures (Fig. 8). Dynamic tests were also undertaken by BRE, using its new laser interferometer, on the new North Stand to provide in-situ data on the fundamental natural frequency and damping ratio of the roof and upper tier. Roof measurements of 1.37 for natural frequency and 0.015 for damping ratio further demonstrated that the new stand roof was likely to be only mildly dynamic and enabled the static wind tunnel test results to be appropriately enhanced for use in design.

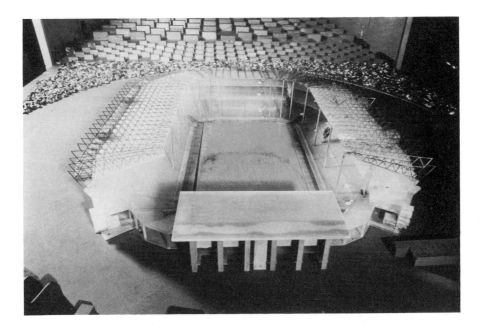

Fig. 8. Wind tunnel testing.

CORROSION PROTECTION

Careful attention was given to both the appearance of the roof members connections and the need to minimise corrosion and maintenance requirements. Where pipe flange joints were used, HSFG bolts were incorporated to clamp the joints together, and any remaining gap was filled and overpainted to ensure a complete seal. All gaps at HSFG load-indicating washers were sealed with a proprietary sealant. Silicone sealant was specified at all forkhead joint details.

Painting, apart from overpainting of bolted connections, was all shop applied to avoid potential problems. Corrosion protection was generally provided by blast cleaning to SA 2.5 and painting with 75 μm minimum DFT of zinc-rich epoxy primer overcoated with two coats of epoxy micaceous iron oxide to 175 μm minimum DFT. Rectangular hollow-section roof purlins were shotblasted and galvanised to 140 μm in recognition of the difficulty presented for their future maintenance.

FIRE ENGINEERING

Fire engineering design was used to assess the required levels of structural fire resistance and, where necessary, to establish suitable fire protection requirements above level 3 where future use of the stand is unlikely to change.

The approach adopted in the original design was broadly in line with the subsequently published British Standard BS 5950: *Structural use of steelwork in building,* Part 8: *Code of practice for fire resistant design.*

Initially, the factors which control and contribute to the temperature rise and burning characteristics of a real fire (such as fire load, ventilation and the materials used in construction) were considered to establish an equivalent period of heating in the BS 476 fire resistance test for all areas of the stands above level 3. An assessment was then made of the fire resistance periods of each structural steelwork member in terms of the BS 476 test, taking into account the position in relation to the fire, the geometry of the profiles, the structural type (e.g. column, beam, strut or tie) and the load ratios (i.e. applied load at fire limit state ÷ load capacity at 20°C).

Fig. 9. Level 4 concourse showing non fire-protected steelwork.

In a substantial number of areas the main structural steel framework was found to have a sufficient intrinsic fire resistance to withstand a real fire without protection.

Fig. 9 illustrates an area of non-fire-protected steelwork to the level 4 concourse. The fire engineering analysis confirmed that the potential fire hazards were concentrated in refreshment bars, debenture holder areas and hospitality boxes, where 60 min fire protection was specified to provide protection well in excess of the calculated time equivalent periods.

The proposals based on this fire engineering approach met with the approval of the London Borough of Richmond-Upon-Thames Building Control Division and were independently assessed by fire engineering specialist Dr B.R. Kirby of British Steel Swinden Research Laboratories, Rotherham.

ENVIRONMENTAL EFFECTS

The importance of minimising any adverse effects on the condition of the pitch was recognised at an early stage. Although this aspect is not always predictable and there are many instances at other grounds where pitches have deteriorated following redevelopment, the likely effects on the quality of pitch needed to be identified.

Specific wind tunnel tests were therefore undertaken at the BRE laboratories to determine the effects of the redevelopment on air circulation at playing surface level. These involved the use of flow visualisation techniques to establish local wind flow directions and areas of high and low winds speeds on the pitch surface and to determine the wind flow variations as each phase of the development was completed. These tests, in conjunction with assistance in their assessment from the Sports Turf Research Institute, indicated there would be sufficient wind flow through the ground after construction of all three new stands to keep the grass well ventilated and healthy. In addition, limiting the roof cantilever to 39 m and the provision of translucent roof sheeting both contribute to minimising the roof shadow and thereby reduce the adverse effects on the pitch growing conditions. In this context the absence of any roof area in the southeast and southwest corners is also beneficial from the point of view of both the wind flow characteristics and avoiding shadow effects.

CONSTRUCTION

The construction of the North Stand was generally advanced from east to west and based upon completing sections of the stand for early use by spectators. The lower tier for example was constructed in six months during the summer of 1989 and handed over for the first match of the 1989/90 season in November. A similar approach was used for the construction of the East and West Stands, advancing construction from the completed North Stand. As each stand was built more structure or increased facilities were required to be constructed in the same time frames. All the partial handover dates were achieved, enabling the RFU to maximise the revenue from spectators during the construction phase.

This achievement was the result of careful planning and exceptional effort by the contractor, combined with genuine teamwork by not only the client, designer and contractor, but also the sub-contractors, local authority and safety team (Fig. 10).

Fig. 10. Model of the redeveloped Stadium.

ASSOCIATED COMMERCIAL DEVELOPMENT

As well as the usual toilet and refreshment facilities for spectators provided along each of the concourses to each tier, there are a significant number of associated facilities embodied within the undercroft of the stands aimed at maximising the commercial success of Twickenham.

The facilities include hospitality boxes, shops, bars, restaurants, fastfood outlets, museum and national fitness centres, as well as changing rooms and a medical centre for the players and match officials.

CONSTRUCTION COSTS AND PROCUREMENT

The North Stand construction was awarded to Mowlem (Southern) Civil Engineering Ltd, based on conventional competitive tender under ICE Conditions of Contract 5th Edition. The construction cost reflected the very buoyant market conditions prevailing in 1989 when both labour and material costs were at a premium in the southeast (Table 1).

The East Stand was also awarded to Mowlem, based upon competitive tendering and on a guaranteed maximum price basis. The market conditions in 1992 were, of course, depressed and this was reflected in the construction cost and the contractor's willingness to accept a guaranteed maximum price approach.

The West Stand was similarly awarded to Mowlem on a guaranteed maximum price basis but, on this occasion, the contract was awarded by single tender negotiation.

Table 1. Construction costs.

Phase	Construction period	Approx. value	Approx. capacity
North Stand	May 89 to Nov 90	£15M	15,000
East Stand	May 92 to Nov 93	£13M	25,000
West Stand	May 94 to Nov 95	£15M	25,000
Associated commercial development	Jan 94 to Nov 97	circa £30M	N/A

The construction costs/seat under each contract are not directly comparable as the extent of associated facilities varied within each contract and a series of separate fitting out contracts was awarded to Mowlem on a negotiated basis. The value of commercial facilities is indicative of the significant increase in terms of costs per seat, that can be incurred in fully developing major stadia venues.

CONCLUSIONS

The redevelopment of Twickenham illustrated in Fig. 11 has totally transformed the stadium from its previous state. It nevertheless continues to remain a mecca for the rugby enthusiast and a special place to visit during the week as well as on match days. The RFU's vision of Twickenham as an attraction for tourists and visitors as well as spectators – 'The Twickenham Experience' – is already being realised.

Fig. 11. View of the redevloped Stadium from the South Stand.

ACKNOWLEDGEMENTS

Client: Rugby Football Union
Architect: Mott Architecture (Mott MacDonald)
Engineer: Mott MacDonald
Contractor: Mowlem (Southern) Civil Engineering Ltd
Safety Team: Richmond Borough Council

REFERENCES

1. The Scottish Office and Dept. of National Heritage: *Guide to Safety at Sports Grounds*, Stationary Office, 1997.
2. *Appraisal of Sports Grounds*, London, Institution of Structural Engineers, May 1991.
3. Cook, N.J: *The designer's guide to wind loading of building structures*, London, Butterworths 1985.
4. Building Research Establishment: *BRE Digest 346: The Assessment of Wind Loads,* 1989.

30 'TENNIS IN AN ENGLISH GARDEN': THE NEW No. 1 COURT AND BROADCAST CENTRE, ALL ENGLAND LAWN TENNIS AND CROQUET CLUB, WIMBLEDON

D. PIKE and R. REES
Building Design Partnership, London, UK

SUMMARY: In 1992 a masterplan for the Wimbledon site was produced which evolved around the theme of "Tennis in an English Garden". The study presented proposals for a New No. 1 Court, a Broadcast Centre and a new Facilities Building to respond to the increasing scale and complexity of the annual Grand Slam Championship. Phase 1 of the development was opened for the 1997 Championships and Phase 2 will be ready in the Year 2000.
Keywords: Bowl geometry, Broadcast Centre, facilities building, precast concrete, roof structure, stadium, tennis, Wimbledon,.

INTRODUCTION

The All England Lawn Tennis and Croquet Club has been the host to the World's Senior Grand Slam Championship since 1922. In the early 1990s it became clear that a major change to its facilities was required to cope with the increasing scale and complexity of the Championships. In 1992, BDP produced a masterplan to a brief which evolved around the theme of "Tennis in an English Garden". The study resulted in a proposal to construct a New No. 1 Show Court together with a new Broadcast Centre and a new Facilities Building, the latter to be built on the site of the old No. 1 Court.

This new development represents the latest thinking in the design of the facilities for this world renowned sporting event. The New No. 1 Court stadium not only provides excellent viewing and playing conditions, but also houses extensive public catering facilities and private hospitality and back up spaces to support them. This has been achieved with a design which incorporates several innovative features including a two lane tunnel under the site, a Broadcast Centre built into the hillside with a landscaped roof and a steel grid shell stadium roof. A clear ordering of crowd movements within the site has also been achieved together with the separation of the front of house, back of house functions. Phase 1 of the development involved the architectural and engineering challenges of cutting into the existing hillside, the articulations of the seating bowl, bowl geometry and sight lines considerations, people movement and the influence of UK Safety in Sports Grounds Legislation. The landscape design reshaped the hillside to provide picnic terraces and to integrate the new road tunnel, the underground Broadcast Centre and two new show courts.

Stadia, Arenas and Grandstands, edited by P.D. Thompson, J.J.A. Tolloczko and J.N. Clarke.
Published in 1998 by E & FN Spon, 11 New Fetter Lane, London EC4P 4EE, UK.   ISBN: 0 419 24040 3

THE MASTERPLAN

Since 1922 there had been continuous change and growth on the site with the All England Club keeping up with the increasing scale and complexity of the championships until the early 1990s when it became clear that a major change was needed. The other Grand Slam venues in France, Australia and the USA had all been rebuilt with Forest Hills Club in New York and the Kooyong Club in Melbourne providing purpose built complexes. The All England Club wanted to keep the responsibility and remain at Wimbledon but to modernise radically. They had to convince their neighbours and the local authority that they could achieve this goal and so in 1992 appointed BDP to produce a masterplan and seek the permissions necessary for the major new works. The brief evolved around the theme of "Tennis in an English Garden". This expressed the desire to focus on the delightfulness of a day at the championships and emphasise the uniquely English nature of it style. Wimbledon is now the only Grand Slam played on grass in a verdant setting of hedges, ivy, flowers and trees. This and the parkland suburban backdrop contrasts sharply with the hot, flat, white concrete glare of Flushing Meadows, Flinders Park and even Roland Garros.

The technical needs were however considerable. Better tennis watching for more people meant increased and improved seating at all courts. Reducing congestion with more landscape meant an enlarged site area with courts respaced plus the relegation of support facilities to underground volumes. The greatly improved broadcast facilities required a large permanent building.

Character was the key goal. The ideal ambience only existed to the South of the Centre Court. This ambience had to be captured and extended over the whole of the enlarged site.

The masterplan (see Fig. 1) is based on Centre Court becoming more central with an equal number of new seats being created to the North. The existing No. 1 Court has been disentangled from the Centre Court complex and housed in a new stadium in the

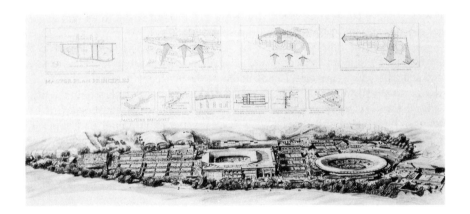

Fig 1. The Masterplan concept.

area of Aorangi Park. In the process, Centre Court has been enlarged and has been given a dignified presence as the 'Country House' in its formal gardens surrounded by parkland.

The extension of the grounds up to the north-western hillside gives a new dimension to Wimbledon, being laid out to allow key views across the site. In addition to the intense enclosures of the courts area, there is now a picnic ground with wide views over London.

Circulation for visitors will eventually become a simple ladder arrangement. The Eastern tea lawn has been extended the full length of Church Road and a new St Mary's Walk will be driven from the hilltop South through the old No. 1 Court location. It follows the vista to St Mary's Church tower as an orientation guide.

The most radical proposal has been to use the landform to create an underground backstage area to the West of the site. The site rises four storeys from Centre Court to the north-west and this level change has been used to build an underground link road across the site and to serve a new central delivery area below the New No. 1 Court. Underground buggy routes link it with all the prime facilities giving secure routes for officials and for players. Ground area is thus released and the task of servicing the buildings can continue uninterrupted throughout the day rather than stopping when the gates open.

Phase 1 of the masterplan has now been implemented and includes the construction of the New No. 1 Court stadium, the Broadcast Centre, the creation of two new show Courts 18 and 19 and the relocation of Courts 14 and 15 which were all opened by the Duke of Kent at the beginning of the 1997 Championship. The construction of the Facilities Building as Phase 2 is now underway and is due to be opened for the 2000 Championships.

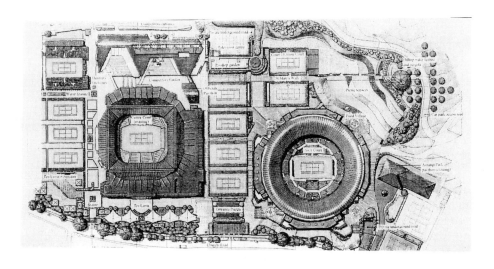

Fig 2. Masterplan.

THE DEVELOPMENT – PHASE 1

The New No. 1 Court

The design brief for the New No. 1 Court evolved and was required to include a varied mix of accommodation. The park in which it was to be located contained thousands of square metres of temporary facilities for the public such as restaurants, merchandising and tickets as well as hospitality suites. These had to be "hoovered-up" and relocated under the new stadium.

In line with other Grand Slam venues, the new No. 1 Court needed to have over 10,000 seats (the existing No. 1 had only 7,000). The final number of seats was just over 11,000 contained effectively into a single tier or bowl. Keeping this number of seats within the desired 60 m of play was extremely challenging with all sides of the Court needing to be maximised for seating. The critical aspect of the seating section occurs at the ends of the Court where sightlines are potentially compromised by the 2 m-high endwall to the Court. This effectively pushes the end tier up. This section then defined the rest of the bowl but with an added lower tier to the long sides of the Court.

Adjustments to the plan geometry achieved excellent sightlines by creating a perfect 12-sided plan with curved linking pieces. This allowed a circular roof to be placed on top of the polygonal bowl without difficulty. The stepped section accommodation blocks tucked below the bowl have been designed to respond to the 12-sided geometry as well as to align them with the rectangular geometry of surrounding Courts (Fig. 3).

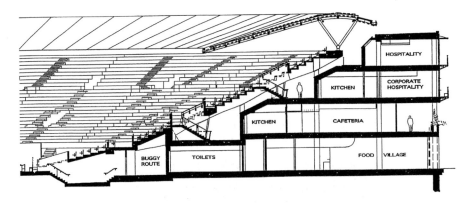

Fig 3. Typical East–West section.

The section provides clear articulation of the building functions as it separates the accommodation and seating bowl. This separation is quite unique in British Stadia (and rare internationally) and creates an exciting and legible space instead of what is often a warren of enclosed passageways.

The buildings have three main structural components; the in-situ concrete slabs and frame to the accommodation blocks; the in-situ concrete raker beams and precast concrete units forming the seating bowl and finally, the steel roof structure.

The first two concrete structures share the same columns. The radial grid of circular columns supporting the seating bowl penetrated the accommodation blocks and also support their flat slabs. Extensive excavation into the clay hillside was required to create a building low enough to satisfy planning constraints. This lead to a piled retaining wall and foundation solution from which the columns rise and which helped to anchor the building into the hillside and prevent clay heave. The actual tennis court was laid on a large concrete tray to prevent differential heave under the grass court.

The upper part of the bowl has around 50 types of precast units totalling some 1000 elements laid on top of the in-situ frame. A shaped in-situ concrete ring beam at the top edge of the bowl takes up the change from the 12-sided seating structure to circular roof.

An unusual part of the brief stated that the shadow cast on the Court by the roof should arrive on the playing surface no earlier than that on Centre Court, where a shadow line clips the tramlines at 5.40 pm on the middle Saturday of the tournament. This requirement together with the sightline criteria determined the height and projection of the roof. The degree of shelter is similar to that on Centre Court, but because of the ability of winter sun to pass under the roof edge grass growing conditions are more favourable.

The roof works as a grid shell structure which is pre-stressed by its own weight with the inner ring members in tension and the outer rings in compression. The overall shape is like a circular dish, supported at its outer edge, with a hole in the middle; approximately 100 m in external diameter with an opening of 70 m diameter. The primary steelwork is all constructed in British Steel's circular hollow section with the tubular roof structure weighing 410 tonnes and the roof in total around 500 tonnes. The lightweight structure allows completely unobstructed views for spectators. The roof is supported on its outer edge by double inclined steel columns which are connected by pin joints to the external circular concrete ring beam (Fig. 4).

Fig 4. Computer image of roof.

Computer analysis made it possible to quickly optimise the structural solution, and as the design progressed the cost was minimised. By minimising the number of joints and by making the most effective use of the steel circular hollow section, it was possible to remove some members. The resultant design is a grid that roughly remains in squares, reducing in size as the stresses get higher towards the inner rim of the roof.

Standardisation and repetition with 72 identical segments to the roof kept the cost down. The total self-weight was also crucial, and a Hoogovens single-ply profiled coated aluminium cladding system was used that tapers and curves in the same way as the structure. Glazing is limited to a strip at the roof outer edge over the accommodation blocks.

The relatively shallow section of the roof gives the impression that it is floating which combines well with the upward views possible from the circulation spaces between the accommodation 'wings' and the bowl (Fig. 5).

Fig 5. New No. 1 Court – Championships 1997.

Designing a roof which inclines towards the Court leads to the challenge of rainwater disposal – i.e. water needs to flow uphill from the gutter. The Centre Court already utilises a system from Japan (Iseki) to force this gravity-defying flow so there was a prototype for the new No. 1 Court roof. It is not a siphonic system but uses a large sealed tank, housed remotely, which has water pumped out of it to create a vacuum. When the water in the gutter reaches a certain level, it is simply sucked away via sealed drainage pipes.

The circulation pattern for spectators is dominated by the need to move them in and out quickly at change-of-ends (the 90 second period where tennis players take a break every two games). Thus there are 40 public vomitories, far in excess of the requirements under the Green Guide. The daylit public circulation space around the bowl functions on only two levels and both levels utilise the slope of the site to allow direct access at grade. The lower level also provides flat access to 40 wheelchair spaces located above the Court level, thus making this one of the most disabled-friendly stadia recently constructed. Four major corner entrances provide access to and from the rest of the site and help to define the four accommodation blocks.

These blocks contain the restaurant facilities that used to be erected each year for The Championships and now have a permanent home under the terraces of the stadium, with views out across the grounds and the adjoining golf course. There are five public food outlets, including the silver service Wingfield Restaurant. Above the restaurants are 11 hospitality suites and the new No. 1 Court Debenture Holders' lounge, overlooking Courts

14 to 17 and across Wimbledon Park to the London skyline. This plethora of restaurants have created a heavily serviced building and given rise to a large number of kitchens and toilets. There are 67 toilet locations altogether. It is astonishing that the whole of this development has been created for use by the Club for only two weeks of the year. The massive 15-lorry underground service yard accessed from a two lane tunnel and basement back-of house facilities are similarly only utilised for The Championships fortnight.

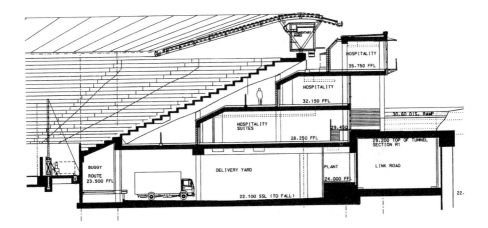

Fig 6. Section at Service Yard and Tunnel.

The Broadcast Centre

Television is vastly important to Wimbledon's fame and to its income. The previous temporary facilities have been replaced by a new permanent building which extends west of Court 14 and the new Court 18. Presentation studios are supported by two underground levels of technical space and a concealed yard for mobile outside broadcast vehicles which is located on the roof of the technical spaces. The host broadcaster, currently the BBC, has the principal corner studio.

The three-level arrangement allows the television companies to de-rig their production support equipment onto the levels below the studio modules and to connect them back to the mobile units in the yard. The lowest level also provides a restaurant for television staff connected to and serviced by the kitchens below the New No. 1 Court (Fig. 7).

The structure and the servicing strategy for the building have been designed to cope with the miles of cabling that are installed in the two weeks before the Championship and de-rigged in the two weeks after. All temporary cable routes are mounted on specially designed overhead cable trays in the corridors and all vertical cabling between production suites and studios is in predetermined risers also on either side of the corridors. The temporary servicing is coordinated with the normal building servicing which has been designed to cope with the heavy cooling loads produced by the recording studios and production units.

The roofs of the recording studios are landscaped as an extension to the hillside picnic terraces and courtside seating has been incorporated into the Northern wing for the new show Court 18.

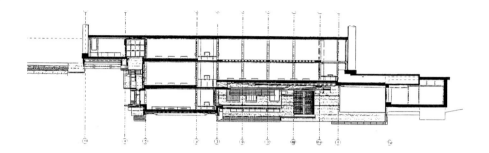

Fig 7. Section at Broadcast Centre.

THE 1997 CHAMPIONSHIPS

At 2.00pm on Monday 23rd June 1997, an almost full house of 11,000 spectators watched Tim Henman beat Daniel Nestor. Many people commented on the intimacy of the setting. The Club wrote "throughout the last month we have heard almost nothing but praise for the new Stadium and Broadcast Centre".

Fig 8. The 1997 Championships.

PART SIX

SUCCESSFUL STADIA PROJECTS AROUND THE WORLD

31 ATLANTA OLYMPICS: THE BIG STORY

R. LARSON and T. STALEY
O'Brien Kreitzberg, USA

SUMMARY The 1996 Olympic Games were held in Atlanta, Georgia, USA, from 19 July to 5 August 1996. The Para-Olympic Games followed 18–25 August 1996. The Games hosted over 10,000 athletes from 197 delegations, more than two million visitors and 15,000 media personnel. Over 3.5 billion people around the world viewed the Games on television. These Centennial Games were the largest Olympic Games in history and were privately funded with a budget of $US 1.8 billion. The Games are estimated to have a economic impact on the Atlanta area of almost $US 4 billion. O'Brien-Kreitzberg Inc joint-ventured to provide program management, design, design management and procurement management for the temporary and portable facilities and equipment (TPFE) for the Games. The TPFE program totalled almost $US 100 million. The Olympic Stadium was constructed to host the opening and closing ceremonies and the athletic events. The budget for the Stadium was $US 220 million. It was designed to be converted to a baseball stadium for the Atlanta Braves following the Games.
Keywords: Atlanta Olympics, baseball stadium, construction management, project management, stadium design, temporary stands, TPFE.

INTRODUCTION

Staging the Olympic Games and the Paralympic Games is different from any other special event in the world. It encompasses a spirit of peace, cooperation and competition unlike any other experience in modern history. The culture of the Olympic movement also implies that the Games are set above the commercialism of privatised events. Although there were published stories of technology, transportation and marketing problems during the Atlanta Games in 1996, the incidents were few and isolated. Overall the Games went very well and the Olympic Stadium operated without a major problem. The conversion of the 85,000 seat stadium to a 47,000 seat baseball venue with a full array of food and entertainment for the entire family was accomplished in less than six months. It took 72,000 staff and volunteers to stage the Games.

The 1996 Centennial Games in Atlanta were an opportunity of a lifetime and the experience of managing large, complex projects to definitive budgets and exacting time schedules cannot be overstated. O'Brien-Kreitzberg Inc (OBK), a wholly owned subsidiary

Stadia, Arenas and Grandstands, edited by P.D. Thompson, J.J.A. Tolloczko and J.N. Clarke.
Published in 1998 by E & FN Spon, 11 New Fetter Lane, London EC4P 4EE, UK.   ISBN: 0 419 24040 3

of Dames and Moore, joint ventured to provide program management, design, design management and procurement management for the temporary and portable facilities and equipment (TPFE) for the Games.

GENERAL

The construction management policy was developed from two over-riding principles – financial prudence, as evidenced by seeking competitive bidding and inclusion, as evidenced by the commitment of the Atlanta Committee for the Olympic Games (ACOG) to equal opportunity and outreach to the community. Overall facility functionality was quite good given the conditions imposed by the ACOG and the authority and responsibility limitations of the facilities and sites. The concerns of spectator population capacity and access control 'throughput' were effectively addressed by 'eleventh hour' site modifications and games-times operational adjustments. Scheduling impacts and spectator injuries were minimal and our personal experience indicated relatively safe and effective crowd flow and accessibility. Some excessive waiting periods (30–90 minutes) occurred at several venue sites. The waits observed were at the security check-in points and at bus shuttle drop-off and pick-up points. Overall the contract responsibilities of the OBK joint venture were performed well with no back-charges from the client and no shortfalls in the Olympic Games.

Publicized pre-event concerns regarding venue accessibility relative to the requirements of the Americans with Disabilities Act (ADA) were not realized and, with minor exceptions due to existing facility conditions, venue accessibility was publicly praised. In order for facilities to be utilised for the Paralympics, additional modifications to accommodate more stringent accessibility requirements were relatively minor. The accessibility provisions of the ADA for the Olympic Games were never taxed and were generally utilized at 10–20% of capacity. However, a 25–30% cost increase should be planned for a facility that complies fully with the ADA.

The pervasive mentality of the ACOG construction and the various contractors was to view the Olympic TPFE program as a short-term special event on a very large scale as opposed to a hard construction project. There are some valid arguments in favour of this view. Due to the scale, visibility, risk and absolute schedule and cost requirements of this event, the classic 'construction project view' was the more pervasive one. We believe we fulfilled the traditional responsibilities of program manager, designer and procurement manager to the best of our ability, enhancing the event and safe attendance by all constituent groups.

Design and construction services for adaptations to existing facilities

Several existing buildings, arenas and auditoriums were adapted to Olympic sports venues through design of minor construction modifications and/or installing temporary equipment and systems. Chief amongst these were the 73,000 seat Georgia Dome, the Georgia World Congress Center, the Georgia Institute of Technology Coliseum, the Omni (15,000 seat city arena) and an airport hanger. The hanger became the Games' Welcome Center at Atlanta Hartsfield International Airport for athletes and officials.

Of the functions criticised by the media (transportation, technology, etc) where we did interact, we believe our design documentation worked smoothly particularly in conjunction with the Olympic Transportation System at those points where they interfaced, and facilitated the functionality of the Technology and Media information systems. We believe the tight construction schedule at many venues and insufficient co-ordination resulted in inadequate pre-event system checkout and 'de-bugging', thus contributing to the games-time performance problems which occurred in the results system.

The decision was made by ACOG to contract with a building contractor on a 'cost plus' basis (for general contractor/construction management of the adaptation of existing venue facilities) on a near staff extension basis. This gave a wide latitude to the contractor for policing its own work. Numerous modifications to design documents were incorporated as the work proceeded. However, these decisions did not result in costly personal injury, litigation or damage to existing permanent construction

The Joint Venture design function's active participation on the project was virtually eliminated from the client's perspective after completion of final construction documents necessary for building permit acquisitions, etc. at each venue.

Temporary and portable facilities and equipment (TPFE)

TPFE commodities or temporary equipment were utilised at 26 sporting venues as well as the Athletes' Village and Airport Welcome Center. One of the critical criteria for the selection of commodity contractors was the consistency/continuity of design and quality for the Olympic Venues. The 1996 Olympic Games was the largest implementation and utilization of temporary and portable facilities and equipment for any event in history.

The TPFE was leased rather than purchased for two reasons: it was cheaper than putting in place permanent facilities to support the millions of Olympic spectators and it made more economic sense for the post-Games owners of facilities. The aquatic venue, for example, seated 15,000 people at the main pool during the Olympics, with all of the necessary supporting concession and toilet facilities. After the Olympics, however, the Georgia Institute of Technology, which took ownership of the facility for university swimming competitions, needed a seating capacity of only about 2,000. Anything more would have been an undue financial and maintenance burden. Each of the 26 sporting venues were temporarily converted into stadia ranging from the 85,000 Olympic Stadium to the badminton competition held in the 2,500 seat Physical Education Building at Georgia State University.

The huge scope of the TPFE can best be appreciated by the following figures:

- 180,000 temporary seats (30,000 of which were in the Olympic Stadium)
- almost 1 million square feet of marquees (tents)
- 25 megawatts of power generators
- 4,000 tons of portable air conditioners
- 200,000 linear feet of temporary fences
- 2,000 portable toilets
- 550 trailers (office caravans)
- 3,000 press and commentator tables
- 3,000 square feet of temporary camera platforms
- supplemental broadcast lighting at indoor venues.

There was a very short mobilisation time for the project. ACOG expected deliverables within two weeks of contract initiation (Notice to Proceed). This expectation resulted in products and processes that in some cases were less than thorough and complete. A reasonable start-up or mobilisation period is required for any project and the length of the mobilisation period will depend on the scope of the project. Prior to the procurement phase for commodities, generic specifications were developed. These specifications were compiled by consultants and staff of ACOG. This caused some confusion to contractors who were asked to provide design/build services. The result of this was that some contractual issues became compromised, causing somewhat liberal interpretations of contracts by all parties. Such confusion caused varying aspects of the contracts to be difficult to enforce.

Requests for Proposals (RFPs) were sent to companies with industry reputations for quality performance. No formal pre-qualification process was used. The return dates for bidders tended to be unrealistically abbreviated. The downside to this was twofold: many qualified contractors refused to bid for fear of providing hurried, erroneous or incomplete proposals. Those who did bid were forced to provide detailed information while not having adequate time to research and prepare. In some instances, ACOG was not provided services that might have been attainable in a better planned scenario and was forced to increase contracts via change orders to accommodate needs. In other cases, the contractors, being ill-prepared, made commitments which were unrealistic and caused additional expenditure on their part to fulfill. Pre-qualified contractors and longer preparation time to contractors would have provided more balanced competition for the contracts and more thorough implementation by the winning contractors.

Since all the TPFE contracts were rentals, it was quite cumbersome to utilise the standard construction contract format. There were often conflicts between requirements of the rental and the general conditions in the contracts. Proper utilisation of a long form rental or service agreement could have streamlined the process and allowed for more clear direction to the contractor.

The overall procurement process was successful. TPFE installations were generally completed on time and the entire project came in under budget. OBK/Russell managed the commodities contractors adeptly and was able to secure positive results.

Documentation of both internal and external issues proved to be an invaluable tool in ensuring the success of the project. The client saw the joint venture as its primary source of information with regard to all phases of the TPFE process. The TPFE contractors were managed deftly and we were able to avoid problems through proactive problem solving and good communication with the construction management staff.

OLYMPIC STADIUM

In July, August and September of 1996 the world's attention was focused on the Summer Olympic Games and the new Olympic Stadium. This state-of-the-art structure was the very first and the very last image for many millions of viewers who watched the opening and closing ceremonies, as well as the athletics events (i.e. track and field). The Stadium was also uniquely designed to accommodate post-Olympic use as the home of the Atlanta Braves professional baseball team.

This dual purpose design approach was the result of one of the primary objectives of ACOG. In planning the Olympic venues, ACOG provided, whenever possible, sites and structures that could be widely utilised beyond the 19 day event, thereby bequeathing an impressive legacy for the benefit of the City of Atlanta, the State of Georgia and its people.

Since the Stadium was ultimately to be the permanent home for the Atlanta Braves, their specific needs dictated the overall architectural design. Thus the outline requirements for the Olympics and for the Braves were identified from the outset, as illustrated below:

Olympic requirements	Braves requirements
85,000 seats	47,000 seats
Field of Play to accommodate IAAF track and field requirements	Field of Play to accommodate all professional baseball requirements
Segregated doping control of athletes	Doping control to be accommodated in locker rooms
Athletes warm-up area and changing rooms	Players locker rooms (home and visitor)
Security for VIPs, Heads of State, Athletes	Security for players
Broadcast camera positions (approx. 25), written press (600) and media facilities	Broadcast camera facilities (approx. 15), written press (50) and media facilities

Design/construction strategy

The design and construction of the Olympic Stadium was approached from a traditional strategy, in that both the designer and the contractor would be contracted directly with ACOG. Preliminary cost estimates were performed by the stadium management team, which established an overall budget of $210 million. This figure encompassed the construction of the Olympic Stadium, re-configuration of the stadium to the final baseball design, land acquisition and construction for 2,500 parking spaces, demolition of the existing Braves stadium, and construction of an athletics warm-up facility for the Olympics at an adjoining site.

Designer selection occurred in early 1992, with initial Requests for Proposals (RFPs) received numbering more than 100. These were subsequently shortlisted to five, and detailed discussions and interviews were held in order to make the final selection. Of paramount importance in the selection criteria, notwithstanding the Olympic International Amateur Athletic Federation (IAAF) requirements, was that the designers had evidenced their understanding of requirements of professional baseball, particularly those of the Atlanta Braves. It should also be noted that inherent with any new organising committee for the Olympics, the criteria and requirements for the Games are continually re-invented to a large extent, particularly as they relate to technology (timing and scoring, broadcasting cameras and infrastructure, video boards, technology infrastructure and integration etc.). The selected designer would need to include adequate specialist consultants to ensure overall stadium design co-ordination of the various elements, which was critical to the Olympics.

Contractor selection also occurred in early 1992, with initial RFPs received numbering over 40. These were shortlisted to three. Here, selection was based upon the contractor evidencing strong stadium construction experience, partnering, value engineering and particularly an understanding of the local market and the Atlanta Braves. The selected contractor was then contracted to ACOG under a Construction Management/General Contractor (CM/GC) Agreement, wherein they acted as construction manager during the schematic design phase of the project for a fixed fee, and provided costs and schedules for the Stadium which they would then sign on to an Initial Guaranteed Maximum Price (IGMIP) contract. Upon execution of the IGMIP, the Contractor continued to provide value engineering solutions to the design development until the overall construction price was within the original budget, and the Final Guaranteed Maximum Price (FGMP) was executed.

Stadium design

From the beginning of the design development phase, the Atlanta Braves management team established the criteria for the theme of the Stadium, and also figured significantly in the space planning. The basis of the design was dominated by the baseball configuration, as evidenced by the 'elbow' in the Southwest corner, which is atypical of the oval track and field stadium design. The Olympic aspect would be incorporated in such a way as to be as temporary in nature as possible, while still meeting all IAAF and Olympic requirements. The design of the Stadium comprised of six levels with the following functional areas per level:

Level	Functional area
Service level	Athletes and VIP drop off areas; main kitchens; main plant and equipment rooms; athletes changing rooms; security and stores
Plaza level	Spectator access to main seating tier through vomitories; merchandise outlets; concessions and toilets
Main level	Served main seating tier; merchandise outlets; concessions; and toilets
Press level	Accommodated written press and media; bar and lounge for same along with audio and video control and editing suites
Club level	Accommodated 60 executive suites; merchandise outlets; bars; and access to 8,800 club seats on middle seating tier
Upper level	Served upper seating tier; merchandise outlets; concessions; and toilets

As the Stadium design developed, with the Braves' requirements outlined in general arrangement drawings, ACOG regularly reviewed their operational needs and whenever possible incorporated permanent areas (i.e. press facilities; concessions; merchandising) for their use without any alterations. However, in many instances the Olympic requirements were far greater in actual space needs. In areas such as the changing room facilities for example, the Braves players' requirements were drastically different from those of the Olympic track and field athletes. To solve this problem, the baseball changing rooms were not constructed and, except for utility supplies, the space was fitted out temporarily for the Olympics by utilizing either metal stud and drywall construction, or drapes and curtains for providing partitioning. This approach was applied in other areas by providing temporary surface mounted power/voice/data lines for the Olympic needs, and through reducing ceilings in areas where an acoustic or aesthetic requirement were not necessary for the function. Such an approach saved costs for the overall project and also added flexibility for space needs for last minute changes.

Construction

The overall construction duration for the Olympic Stadium was 30 months from groundbreaking to the final hand-over to ACOG (approximately four weeks prior to the Opening

Ceremony). The construction was fast-tracked in order to meet the time frame, however, the true critical element in the schedule was the issuance of completed design drawings.

The main components of the stadium structure consisted of 2,650 bored piles which were installed in groups of four to twelve with a reinforced concrete pile cap. Structural grade beams tied between pile caps, and structural slab on grade capable of HA loading was installed for the whole Service level. In-situ reinforced concrete columns from the Service level to Plaza level, with a reinforced concrete slab completed the Plaza level. From the Plaza through the Upper level and the Roof canopy, structural steel with metal decking and wire mesh reinforced concrete completed the superstructure. From the Field of Play to the Main level, the seating bowl consisted of precast raker beams and triple riser precast seating terraces. Above the Main level, the precast seating terraces were placed on structural steel cantilevered rakers.

All of the permanent precast seating terraces were applied with a backer rod and water-proofing membrane on all horizontal and vertical joints. This was crucial as the concourses on the underside were mostly operational areas such as concessions, merchandise or toilet areas. Additionally, as the Stadium was open air, all concourses included a waterproofing membrane within a sandwich slab concrete construction.

The scope of the Stadium construction can best be appreciated by the following figures:

- 2,650 bored piles/159,000 lineal feet at 65 linear feet each
- 55,000 cubic meters of earth excavation
- 65,000 cubic meters of concrete
- 8,500 tons of structural steel
- 3,100 tons of reinforcing steel
- 1,600 pieces of architectural precast concrete
- 15,000 square meters masonry
- 2,000 square meters of aluminium curtain wall
- 2,000 square meters of metal roof decking
- 12 miles of handrail
- 7,059 light fixtures
- 401 high output speakers mounted on roof
- 749 low output speakers in concourses, suites, toilets.

Partnering, cooperation, patience and communication from all the team members throughout the numerous difficult times was absolutely vital to the successful completion of the stadium project. Knowing that the eyes of the world would be focused on the Olympic Stadium served as both high pressure and high incentive to ensure that the best possible stadium was presented during the Summer Olympic Games. And after all the grand celebrations of the closing ceremonies was over, it was the dedication of the management, designers and contractors and all others involved in the immense undertaking who then completed the difficult conversion, and produced a world class baseball stadium.

32 HIGH TECHNOLOGY STADIUM CONSTRUCTION BY TAKENAKA CORPORATION, JAPAN

M. MIYAMOTO
New Frontier Engineering Department, Takenaka, Tokyo, Japan

SUMMARY: An outline of the Takenaka Corporation introduces the paper. Plans for the 2002 Football World Cup to be played in ten cities in Japan are then described. The main part of the paper reviews the new stadia that are being planned and built in preparation for the series of matches. The stadia include a wide range of facilities, and are designed to meet many local conditions, and with many innovative structural and construction solutions. These include dealing with difficult ground conditions, sophisticated designs of roof structures, design of stadia for use for both football and baseball, moveable seating, and ways to allow natural turf to be provided which can be used in covered stadia. At Yokohama, this involves the use of a complex air-supported structure which moves the entire playing field, weighing 8000 tons.
Keywords: Air-supported structures, dome-shaped roofs, flood risk, Football World Cup 2002, Japan, movable seating, movable sports fields, precast concrete, pre-stressed concrete, retractable roofs, roof covering, turf grass.

TAKENAKA CORPORATION

The Takenaka Corporation is one of the 'Big Five' construction companies in Japan, most of which are involved in building and construction works. The work undertaken in 1997 by the company amounted to approximately five billion pounds, excluding civil engineering works. The civil engineering works are executed by Takenaka Civil Engineering & Construction Co., Ltd, an affiliated company.

About 90% of the work undertaken by Takenaka Corporation is within Japan, but the company has expanded overseas too, and has 31 overseas offices today. In the United Kingdom, Takenaka U.K. Ltd has offices in London, Cardiff and Birmingham.

A key feature that sets Takenaka Corporation apart from others is its modus operandi, based on the concept "Trust is everything. Discharge your responsibility in good faith and win total customer satisfaction."

Another distinctive feature of Takenaka Corporation is that the company adopts the ideal of an integrated design/construction system. Naturally, this ideal is based on the premise of 'reliability and sincerity'. The company employs experienced staff to faithfully reproduce the quality desired by its clients through the consulting, design, engineering and construction stages, and delivers 'products' of good quality by adhering to the estimated

Stadia, Arenas and Grandstands, edited by P.D. Thompson, J.J.A. Tolloczko and J.N. Clarke.
Published in 1998 by E & FN Spon, 11 New Fetter Lane, London EC4P 4EE, UK.   ISBN: 0 419 24040 3

cost and delivery period. The results of the company's past achievements indicate the ratio of design/construction works to negotiated contract works as approximately 0.5.

In 1979, the company received the Deming Prize for its Total Quality Control (TQC) activities, the first construction company in Japan to receive this prestigious award. In 1992, Takenaka Corporation received the Japan Quality Control Medal, proving to the world the excellent quality management of the company.

Takenaka Corporation is not satisfied by merely preparing design drawings. Based on its rich experience over the years, the company takes into account factors that are not visible in design drawings and offers detailed proposals with the aim of creating better quality products.

JAPANESE SITES FOR HOSTING FOOTBALL GAMES IN THE JAPAN-SOUTH KOREA 2002 WORLD CUP

The 2002 World Cup will be hosted jointly by Japan and South Korea. Initially, 15 cities in Japan were selected as venues for hosting the football matches, but this list was finally trimmed down to ten cities by the authorities concerned. The cities selected (shown in Fig. 1) are Sapporo, Sendai, Niigata, Kashima, Urawa, Yokohama, Shizuoka, Osaka, Kobe and Oita. As of today, the construction of facilities for hosting the football games are complete only at Yokohama and Osaka; others are in the design stage or in the preparatory stage of construction. Takenaka Corporation is the main builder for the facilities at Yokohama and Osaka; moreover, it has also received orders for design and construction of the facilities at Sapporo and Oita. Detailed specifications for construction of football stadia for hosting World Cup games are prescribed by FIFA. All the stadia at the ten locations are to be newly built or remodelled according to FIFA specifications. Out of the ten stadia, two are to be built specifically as football stadia, seven as stadia for football games and athletics, and one as a dual purpose football/baseball stadium. The stadia at Sapporo and Oita are dome-shaped while the others are open air stadia with a roof above the spectator stands. Of the four stadia whose construction Takenaka Corporation has been entrusted with, the three stadia at Yokohama, Oita, and Sapporo are introduced here.

YOKOHAMA INTERNATIONAL STADIUM

The Yokohama International Stadium (Fig. 2) is Japan's largest athletics-cum-sports stadium with a seating capacity of 70,366, completed in December 1997. It is located on the flood plain of a nearby river: providing a pitch at the usual level involved a risk of flooding in the event of heavy rains. Therefore, the ground was artificially built up to a level of 10 m above the ground using sand and turf, which was planted to be used as pitch.

For building the artificial ground and the spectator stands, a connection method using prestressing of precast reinforced concrete members was adopted. This technique involves manufacture of columns and beams separately, and tying them together using prestressing strands to form rigid frames. By precasting, the quality of the members is improved and prefabrication allows the members to be taken to site and subsequently assembled, thereby shortening the construction period.

The design of the stadium was carried out jointly by two of Japan's largest design

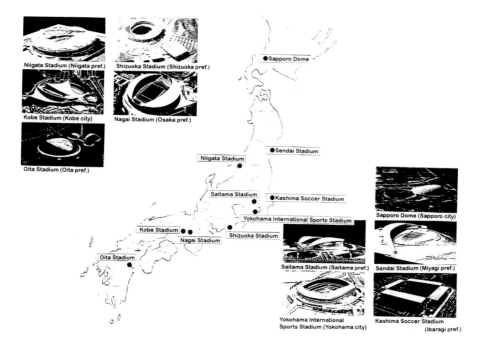

Fig. 1. Sites for the 2002 World Cup in Japan.

Fig. 2. Yokohama International Stadium. Aerial view.

offices, Matsuda Hirata and Touhata, while Takenaka Corporation offered various proposals for the construction methods. One of the proposals was for the construction of a large roof covering two-thirds of the spectator stands. The conventional method employed in Japan is to build a temporary platform, erect the roof on it and then remove the temporary platform. However, if this method is used, work below cannot be started until the roof is completed. Takenaka Corporation proposed the diagonal suspension method,

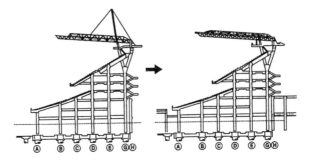

Fig. 3. Yokohama International Stadium: roof structure.

whereby the roof would be lowered by suspending it from erected posts, as shown in Fig. 3. Using this method resulted in shortening the construction period by about two months.

In adopting the precasting method, an enormous number of members totalling 54,000 pieces had to be controlled. To cope with this problem, an information integrated construction system was developed, which included control of inflow of materials, erection control, work progress control, and control of hoists.

Furthermore, to give priority to safety, the large scale composite construction method was adopted, accuracy was improved and efficiency in surveying work was enhanced by developing a three-dimensional survey system. Foolproof construction plans are being prepared covering quality, completion period and safety measures in work.

Yokohama is conveniently located only 20 minutes from Tokyo by the Shinkansen (high speed train). The finals of the World Cup games are likely to be held at the stadium in Yokohama.

OITA STADIUM

The Oita Stadium (Fig. 4) of Oita Prefecture is located in Kyushu, and is about an hour and a half from Tokyo by plane. The most distinctive feature of this stadium is the dome-shaped stadium with hemispherical retractable roof.

Fig. 4. Oita Stadium.

A design/construction competition was held for planning the Oita Stadium, giving top priority to the specifications of FIFA and giving further consideration to its use after the World Cup. Takenaka Corporation formed a consortium with the internationally-renowned Kisho Kurokawa, cleared all hurdles in the given proposition one by one, and prepared a proposal for submission for the competition. Takenaka Corporation was the first candidate to be selected based on its proposal.

The work on the Oita Stadium is expected to commence in June 1998. Currently, the final stage of detail design is in progress. Here is a brief introduction to the Oita Stadium.

The maximum seating capacity of the Oita Stadium is 41,957 (during the World Cup). During athletics competitions, the portable seats will be removed so that the total capacity will become 35,590. The elliptical portion of the steel-framed roof is open and the roof is constructed so that the bow-shaped movable part of the roof slides over rails (Fig. 5). The fixed part of the roof is made of titanium plates, while the movable part of the roof makes use of high strength, high transparency web members with a 25% transparency rate to ensure adequate brightness when the roof is closed.

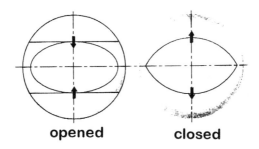

opened   **closed**

Fig. 5. Oita Stadium: roof opening.

Takenaka's Research and Development Institute, based in the prefecture of Chiba, Japan, has been the major support facility for managing the complex logistics involved in the design and construction of stadia. This facility has been used to help deal with many of the key issues involved in the development of the design process, the construction logistics, and any subsequent monitoring after the building has been completed.

This is particularly necessary in the process of designing domed-roof stadia, where issues such as refuge from fires, air conditioning plans, acoustic levels, and illumination systems are a few of the key areas of consideration when catering for very large numbers of people.

These important criteria have been resolved to their optimum, making full use of Takenaka's engineering expertise fostered by its experience in the design and construction of the Tokyo Dome (Fig. 6), the Fukuoka Dome (Fig. 7) and the Nagoya Dome (Fig. 8).

Another problem that arises when the roof is dome shaped, is how best to plant and grow turf. Oita is at a low latitude compared to the rest of Japan, but since the roof is retractable, the problem of preventing direct sunlight from striking the turf does not arise. However, in winter, the shadow of the fixed part of the roof to the south hinders the growth of turf. On the advice on the design of the roof by Dr James B. Beard, it was decided to use the Advanced Turf System based on the Texas USGA method so that a soft but elastic pitch

Fig. 6. Tokyo Dome.

Fig. 7. Fukuoka Dome.

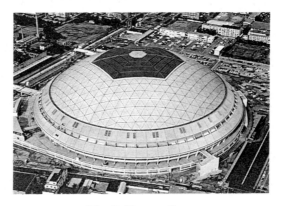

Fig. 8. Nagoya Dome.

can be realised and playing on the pitch is comfortable. The Zoisha grass to be used in the Oita Stadium is likely to wither away in winter, therefore we are contemplating an underground heating system using pipes to prolong the period the turf remains green. In addition, this stadium will be designed for multi-functional use and will incorporate various equipment, such as: sky cameras that travel with and capture a player as the player runs after the ball; a multi-stage system enabling a stage for holding concerts to be installed without adverse effects on the turf; wall curtains to partition the space according to the size of the event; portable seats etc.

The Oita Stadium is scheduled for completion in March 2001, and will be used for pre-World Cup games.

SAPPORO DOME

The Sapporo Dome (Fig. 9) is also a project in which the designer and builder were selected based on the results of a design/construction competition, similar to the Oita Stadium. For this project, Takenaka formed a consortium with Koji Hara as the designer and Taisei Corporation as the builder, submitted a proposal and became the first candidate to be selected.

Sapporo is the city that hosted the 1972 Winter Olympics. In a severe winter, as much 12 m of snow falls in Sapporo. In view of its location in a cold region, a stadium in Sapporo cannot naturally be used for outdoor sports. The citizens of Sapporo eagerly anticipated a large space that can be used in winter for sports. Japan, like the US and Cuba, is a country where baseball is a popular sport and Sapporo has long clamoured for a baseball stadium.

The city of Sapporo presented the topic 'A dome for the World Cup football games and a baseball stadium' to the participants of the design/construction competition.

Fig. 9. Sapporo Dome.

Like the Oita Stadium, the most important of the effects of a dome is its effect on the growth of turf. Another factor that needs consideration is that the arrangements of spectator seats in a football stadium and a baseball stadium are radically different, Figs 10 (a) and (b).

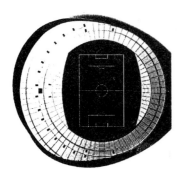
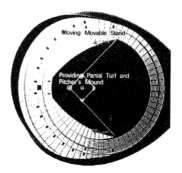

(a) Seating for soccer.   (b) Seating for baseball.

Fig. 10. Sapporo Dome

The solution to both these problems is the concept of a dual arena, where the turf is grown outdoors and brought indoors at the time of play. The field with natural turf measures 86 × 120 m, and the entire field (as a single piece.) is moved indoors when required. The concept of the field as a single piece is a necessary condition, because a football field cannot have joints. This large field is made of a steel framework with concrete floor slabs, sand and turf, and weighs about 8,000 tons. Air pressure is used to move the field. The principle is similar to that used in a hovercraft; we call it the hovering stage. The hovering stage is constructed so that it is capable of lateral as well as rotary movements. To bring in a stage of width 86 m, an opening of at least 86 m in the dome is necessary. However, merely creating an opening of 86 m gives an impression of a gigantic empty space or nothingness when viewed from inside the dome. We have provided an arrangement whereby the spectator stand in the way of the opening can be collapsed, rotated and stowed so that the hovering stage can be passed through the opening and the spectator stand erected again so that the space becomes closed. (Figs 11 and 12)

When using the arena as a football stadium, the stage brought in is rotated further and matched with the main stand to create a seating capacity of 43,000 as required by FIFA standards for a football stadium, Fig. 10 (a). When using the arena as a baseball stadium, the stage is removed and taken out of the stadium, turf is laid out, and the rotatable movable seats are moved into position to create a seating capacity of 42,243, Fig. 10 (b).

The domed roof is a steel structure with the entire roof covered by stainless steel. Similar to the Oita Dome, the various topics related to this dome were analysed under varying conditions, simulations performed and problems resolved. Particularly, special considerations have been given to the location of the stadium in a snow and cold-prone area, snow falling from the roof, maintenance of a comfortable heated environment, the lightweight hovering stage, growth of turf, landscaping considering maintenance of the ecosystem, etc. in the preparation of plans.

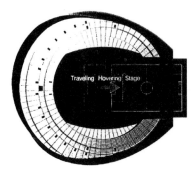

Fig. 11. Sapporo Dome: travelling hovering stage.

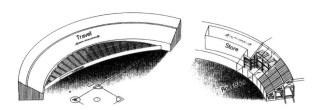

Fig. 12. Sapporo Dome: movable seating.

In recent years, energy savings, maintenance-free structures and reduced life cycle costs are being demanded from the aspects of prevention of global warming of the earth and cleaner environments. The design of the Sapporo Dome proposed by Takenaka Corporation incorporates all these matters.

The Sapporo Dome is scheduled for completion in March 2001, the same time as the completion of the Oita Stadium.

In 2002, a new type of stadium, radically different from the World Cup stadia until now, will make its appearance. We hope that you will all visit Japan and see the World Cup games at the Sapporo Dome.

33 THE AMSTERDAM ARENA: ONE OF THE MOST INNOVATIVE ENTERTAINMENT FACILITIES IN EUROPE

H. J. MARKERINK
Amsterdam Arena, The Netherlands

SUMMARY: The Amsterdam ArenA is a multi-purpose sports and events arena, located in south-east Amsterdam. Construction started in 1993. It is the home to the Ajax Football Club, and is used for many types of sporting and entertainment events.
Keywords: football, multi-purpose stadium, retractable roof.

IMAGE

Our country has, by the realisation of the multi-purpose sports and events Arena in Amsterdam, one of the most innovative entertainment facilities in Europe. A unique feature of the Arena is its retractable roof. The new Arena completely corresponds to the way in which top sports are played in our country.

In the international football world, The Netherlands have an outstanding record: twice runner-up in the World Cup and once, even, European Champion... not to mention the many European Cups won by Ajax.

Within Europe the Arena attracts considerable attention. Well-known artists like Tina Turner and Michael Jackson as well as many sports organisations have already used the Amsterdam ArenA's excellent facilities. Spectators from all over the world are visiting Amsterdam to see and experience the Arena.

CONNECTIONS

The Amsterdam ArenA is located in the south east corner of Amsterdam, near Strandvliet underground station and the underground/railway station at Bijlmer.

Overground and underground trains stop right outside the Amsterdam ArenA. Separate footbridges provide a direct link between the Arena and the station platforms. Highways A2 and A9 provide access to the Arena for those travelling by car. The 'transferium', a large two-storey car park beneath the Arena, provides parking facilities for some 2,600 cars. A further 8,500 parking spaces are available within walking distance of the Arena.

Stadia, Arenas and Grandstands, edited by P.D. Thompson, J.J.A. Tolloczko and J.N. Clarke.
Published in 1998 by E & FN Spon, 11 New Fetter Lane, London EC4P 4EE, UK.   ISBN: 0 419 24040 3

AJAX FOOTBALL CLUB

The Amsterdam ArenA is the home ground of the famous Amsterdam Football Club Ajax. Whenever you mention Amsterdam, people immediately think of Ajax, Johan Cruyff, Marco van Basten and all the other Ajax star players. A club of this stature deserves an Arena with international appeal.

The Amsterdam Arena aims to be associated with Amsterdam in the same way that Wembley is associated with London and Nou Camp with Barcelona.

MULTI-FUNCTIONAL

Besides football, American Football is played in the Amsterdam ArenA as well; the Amsterdam Admirals play their home games here. For example rugby, tennis or any other sport could also be organised in the Amsterdam ArenA. The Arena also hosts major (pop) concerts and mass meetings.

Through an arena-setting smaller events can be organized in the Arena, such as boxing matches, tennis tournaments, basketball competitions, TV shows, carnivals and ice shows. It goes without saying that the other major football competitions, such as national and European Championship finals, will also be played in the Amsterdam ArenA. The restaurants of the Arena are in use on a continuous basis for all kinds of business meetings, conferences, seminars, etc.

COMFORT

In designing the Amsterdam ArenA, considerable emphasis was given to the safety and comfort of the spectators. Every seat in the covered, all-seater Arena has an uninter-rupted view of the field. In the wide walkways around the Arena there are many outlets for food and beverages. And, thanks to the many TV monitors, there is no need for anyone to miss a moment of action on the field. It is a very user-friendly Arena, which gives enormous pleasure to a great many people.

CONSTRUCTION

The construction of the Amsterdam ArenA started in 1993. Commissioned by Amster-dam ArenA/Stadion Amsterdam NV, it is being built by Bouwcombinatie Stadion Amsterdam, a consortium comprising Ballast Nedam Utiliteitsbouw and Bam Bredero Bouw.

The engineering firm Grabowsky & Poort BV was responsible for the Arena's architecture and construction and Twijnstra Gudde NV acted as the representative of Amsterdam ArenA/Stadion Amsterdam NV. The Arena's architect is Rob Schuurman and Sjoerd Soeters designed the entrance building.

FACTS AND FIGURES

The Amsterdam ArenA has:

- A total of 51,000 seats, including:

9	10-seater founders' lounges
54	10-seater sky boxes
200	VIP seats
1,384	business-seats
3,460	seats reserved for shareholders
130	seats reserved for the press
46,000	other seats.

- Two stands: the lower stand providing seats for 23,500 spectators; the upper stand providing seats for 27,500 spectators.

- External dimensions of:

length:	approximately 236 metres
width:	approximately 165 metres
height:	approximately 75 metres.

- 2,200 lux floodlights, radiation heaters and a PA system, all of which are suspended from the Arena's inner roof-structure.

- Retractable roof consisting of two sections, each measuring 35×118 metres, which can be opened or closed in about 30 minutes.

- 60 entrances to the stands.

- Three concourses on three different levels behind the stands.

- 50 outlets for food and beverages at the concourses.

- Ample parking facilities, including: a transferium situated beneath the Arena with capacity for 2,600 cars, plus a further 8,500 parking spaces in the immediate vicinity of the Arena and within easy walking distance.

- Two giant video screens (Nitstars).

- A museum exclusively devoted to Ajax; a TV studio; numerous restaurants (approximately 1,500 m²), conference rooms and, of course, state-of-the-art facilities for the players.

34 DESIGN OF ANYANG CITY GYMNASIUM, KOREA

J. S. KIM, Y. N. KIM and K. M. KIM
C.S Structural Engineers Inc, Seoul, Korea

SUMMARY: The Anyang City gymnasium has 6000 seats for handball, basketball and volleyball. It is designed with a steel truss roof and columns. The roof trusses are covered with sandwich metal panel and the truss columns are exposed to the air. This article shows how the design of the roof and column effectively use the concept of the structure.
Keywords: Lattice columns, Korea, metal sandwich panels, pin base connections, stadia roof, steel roof trusses.

INTRODUCTION

The gymnasium is located in Anyang City near Seoul, Korea (Fig. 1). The Architect is Hang-Lim Architects Inc., Seoul in association with Anderson & OH Inc., an architectural firm in Chicago, Illinois, USA. The structural consultant is C.S Structural Engineers Inc., Seoul in association with Tylk, Gustafson and Associates, Inc., Chicago, Illinois. The contractor is Doosan Construction Inc., Seoul. The construction period is from September 1997 to October 2000.

Fig. 1. Artist's impression of Anyang City gymnasium.

Stadia, Arenas and Grandstands, edited by P.D. Thompson, J.J.A. Tolloczko and J.N. Clarke.
Published in 1998 by E & FN Spon, 11 New Fetter Lane, London EC4P 4EE, UK.   ISBN: 0 419 24040 3

STRUCTURAL DESIGN

Shape

The shape is an elliptical roof supported with 12 trusses, the eaves height is GL + 15. 22 m and the maximum height of roof truss is GL + 24, 33 m (See Figs 2 and 3).

The dimensions of the roof are 94.50 m long and 72.00 m transverse. The roof trusses consist of an inverted triangle pipe truss, 1.30 m deep at the centre and 3.00 m at the eaves end (see Figs 4 and 5).

The supporting column is a triangle lattice column with a pin base end (see Fig. 6).

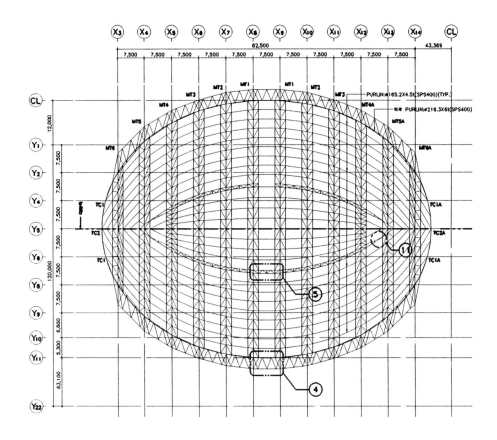

Fig. 2. Roof framing plan.

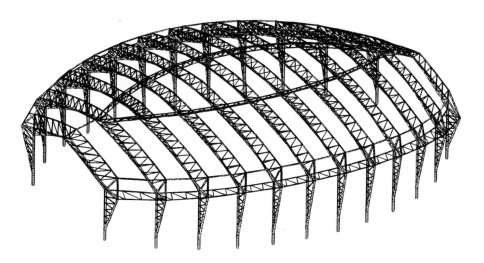

Fig. 3. Roof trusses and lattice columns.

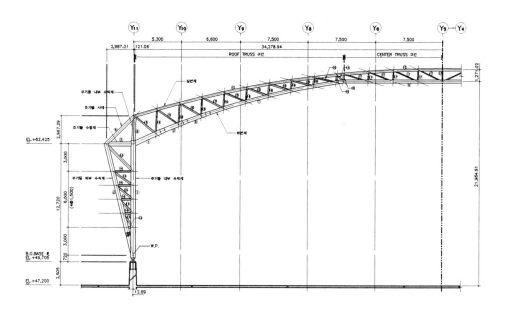

Fig. 4. MT1 truss elevation.

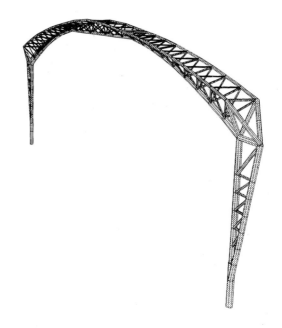

Fig. 5. Perspective of truss.

Analysis and structural system

The dead load (DL), the live load (LL), the wind load (WX, WY), the snow load (SL) and the temperature load (TL) are considered for the analysis and combined for the worst case, as follows:

1. DL + LL
2. 0.76 × (DL + WX)
3. 0.75 × (DL + WY)
4. 0.76 × (DL + SL)
5. 0.75 × (DL + TL)

The portal-shaped one-way trusses carry the roof loads to the lattice columns and foundations in the transverse direction of the ellipse roof. For horizontal stability, the bridging trusses are connected to each portal frame in the middle and the eaves truss are connected to each lattice column at the eave of the roof truss.

Material

The yield strength of pipe truss and lattice pipe column is 3300 kg.f/cm^2 and the strength of purlin and bridging pipe truss is 2400 kg.f/cm^2. The 28-day cylindrical compressive strength of the foundation concrete is 240 kg.f/cm^2.

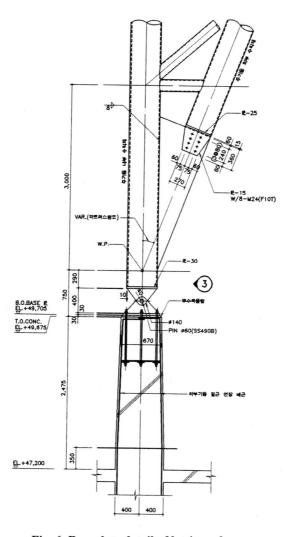

Fig. 6. Baseplate detail of lattice column.

CONCLUSION

The 12 portal truss frame whose lattice columns are exposed to the air are very sophisticated and very effective for supporting the roof. The total quantity of steel including the main frame and purlins is 71 kg/m². The bridging and eaves trusses act for the wind loads and the vertical loads effectively.

REFERENCES

1. The code of steel structures, The Architectural Institute of Korea.
2. Manual of steel construction, 10th edition, AISC.

35 AIR CANADA CENTRE: STRUCTURAL DESIGN PROCESS

D. K. WATSON
Yolles Partnership Inc. Consulting Structural Engineers, Toronto, Canada

SUMMARY: This paper is a presentation of the structural design process for the concrete portion of the Air Canada Centre, Professional Basketball and Hockey Arena. This process took place within the context of a design/build contract, therefore, the interaction with the contractors during the design represents a major portion of this paper. The importance of identifying and resolving the buildability issues during the design is discussed with specific examples being given. These examples will demonstrate that the design must include for the coordination between all the structural sub-trades in order to achieve a buildable design.

This paper also outlines the engineering analysis and design process with respect to the control of complex structural designs. With today's powerful computers, it is becoming commonplace to model complete three-dimensional structures. For a large structure such as an arena, the use of simpler design approaches to check the complex design is imperative. The methods used to do this for the thermal, seismic and gravity designs are outlined below.

Keywords: Arena, structural design, buildability, moment frame, thermal design, seismic design, fast track, design build.

INTRODUCTION

The Air Canada Centre is a multi-purpose facility consisting of an arena and an adjacent office tower. The arena has a seating capacity of approximately 21,000. Its primary uses are for professional basketball and hockey, but it has the flexibility to be used for concerts, theatrical events and conventions. The office tower is a fifteen storey structure, which will be occupied by the sports team administration and Air Canada, a corporate sponsor.

The arena is comprised of a reinforced concrete substructure and superstructure, supporting a long span, structural steel truss roof. The office tower is a reinforced concrete, flat slab structure which is physically separated from the arena with an expansion joint above grade. For the purpose of this paper, only the arena concrete structure will be discussed.

Stadia, Arenas and Grandstands, edited by P.D. Thompson, J.J.A. Tolloczko and J.N. Clarke.
Published in 1998 by E & FN Spon, 11 New Fetter Lane, London EC4P 4EE, UK.   ISBN: 0 419 24040 3

The arena consists of three levels below grade which include two partial levels of parking below the main event level. The event level contains the basketball court/hockey rink, change rooms, private boxes and all main service facilities such as kitchens, shipping and receiving, and storage. The at-grade level is the lower concourse which provides the major access to the facility, along with concessions and washrooms (refer to Fig. 1). Above grade are two main floors, the suite level consisting of private boxes, and the upper concourse which provides access to the upper seating level along with washrooms and concessions (refer to Fig. 2). Above the upper concourse are three loge mezzanines at the west and two balcony mezzanines at the east. Finally, there are two mechanical rooms housed within the roof structure.

In order to ensure that the facility was designed and constructed to the project budget, using practical and buildable techniques, a general contractor was selected during the design process. The consulting team was able to present alternative design solutions to the contractor who in turn assessed these for cost effectiveness and feasibility. This process was iterative and allowed the final design to be established with confidence, eliminating potential delays created by construction problems which occur when buildability issues arise during construction. This paper outlines the design/build process and gives specific examples of issues that arose and how they were solved.

The arena "bowl" is a reinforced concrete slab, beam and column structure which derives its lateral stability from frame action. The building plan dimensions are approximately 105 m x 120 m and have been designed and constructed without expansion joints within the "bowl". The only separations are from the adjacent above-grade office tower.

This structural arrangement requires that care be taken in the lateral, gravity and thermal analyses and designs. This paper outlines the processes used stressing the importance of using simple models and hand calculations to confirm the design results attained from complex computer models.

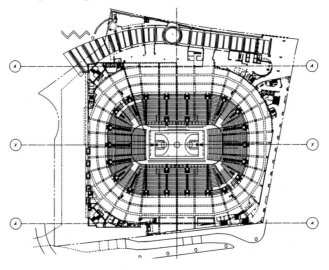

Fig. 1. Plan view of lower concourse level.

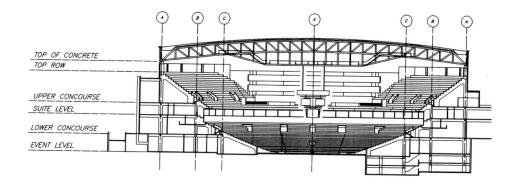

Fig. 2. Building section.

BUILDABILITY ISSUES

Design build process description

The involvement of contractors in the design process is essential for large, complex building projects. To initiate this for the Air Canada Centre, the consulting team prepared a schematic design package of drawings and specifications. PCL Constructors Eastern Ltd. was selected even though their bid was submitted as a budget price. It was their desire to work with the consulting team to refine the design further prior to committing to a guaranteed maximum price.

The refinement of the design started with PCL presenting their first value engineering document, which identified a list of global and detail issues for all the consulting disciplines to assess from a technical perspective. Examples of the global issues affecting the structural design were:

1. Minimization of the basement/event level floor height to reduce excavation and finishing costs.
2. Relocation and rearrangement of the parking levels below grade to reduce excavation and temporary excavation retention adjacent to the TTR rail corridor along the north property line.

Examples of detail issues affecting the structural design were:

1. PCL proposed reducing the depths of floor beams spanning between the main frames, and making these beams wider. The rationale behind this approach was to create a floor framing system which was more like a slab and continuous drop panel

arrangement than a beam and slab system. PCL identified that this would reduce formwork costs. Even though more concrete would be required, this cost was more than offset by the formwork savings.

2. PCL expressed their desire to use a steel formwork system for the main frames. This would allow frame beams to be poured without a shoring system since steel formwork is self-supporting. The decision to proceed with this required a review of the impact on the reinforcing detailing of beam or raker to column joints.

From the above, and other value engineering items, the structural analysis and design was reviewed and updated. The details of these are discussed later in this paper. A design development package of drawings and specifications, including estimated unit quantities where applicable, was prepared and issued for tender to the various structural subtrades. After the selection of subtrades, the design/build process continued. Particular attention was paid to the coordination between the steel formwork and reinforcing steel systems.

Other construction issues were also coordinated during the design/build process. Examples of these are as follows:

1. In order to provide sufficient space for the cranes to be used for erecting the main roof structural, an area larger than the court surface, at the event level, was required. For this reason the seating rakers from the concourse level down to the event level had to be omitted temporarily. It was decided to use precast rakers for this purpose. This resulted in a system which was faster to build and reduced the structural stiffness of the lower level which reduced the thermal restraint and reinforcing quantities at this level.

2. Access to the construction site was very limited due to the proximity of an adjacent rail corridor, elevated Gardiner expressway and city streets. This required the coordination of the structural design for temporary access to the lower event level with portions of the structure omitted temporarily. In addition, the area of structural slab-on-grade immediately to the north of the arena was designed for construction material and concrete truck loading.

Final slab, beam and column system description

As a result of the design/build coordination process, a final slab, beam and column arrangement was selected. The following describes this system and the reasons for selecting each component of the arrangement. Figures 1, 3 and 4 illustrate the framing.

1. The main frame beams indicated on number grid lines are typically 750 mm wide by 900 mm deep. These are supported on 1,000 mm by 600 mm columns. The beams were chosen to be wider than the columns to minimize the reinforcing congestion at the joints. Figure 4 illustrates how the wider beams allow space for flexural steel to fit outside of the column verticals.

2. The secondary beams framing perpendicular to the numbered radial grid lines were selected as 500 mm deep with varying widths. At the beam to column intersections, a width of 900 mm was used which is 100 mm less than the columns. This width maintained the flexural strength and stiffness required for gravity loads and stability.

In addition, the width provided sufficient space for reinforcing steel while aligning this steel between column verticals.

3. All these member dimensions were selected for strength and serviceability requirements and to minimize reinforcement congestion. Columns were sized to achieve a maximum of 2.0 to 2.5% by volume vertical reinforcement. The beam dimensions were chosen to maintain a layer of negative flexural reinforcement for secondary beams and two layers for main frame beams.

4. For all of the above member size selection, a minimum concrete cylinder strength of 30 MPa was used. This was increased to 35 MPa at some heavily loaded columns and for the parking garage columns to conform with the requirements of the Canadian Standards Association standard for parking garages[1].

Standardization of reinforcement details for beam column joints

The construction of a moment frame structure is simplified by the standardization of reinforcement arrangements. To this end, the locations of main flexural reinforcement in beams and vertical reinforcement in columns were kept constant in each type of member. For beams, the maximum number of bars in each layer were kept constant. For columns the number of bars was maintained at 16 with reinforcement quantities adjusted by changing bar size. The result was a very repetitive structure which increased the productivity of the reinforcing steel installers.

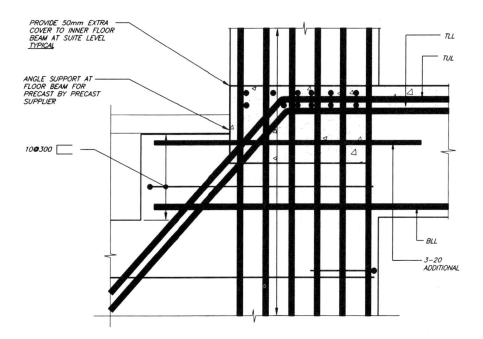

Fig. 3. Section of typical beam to column joint.

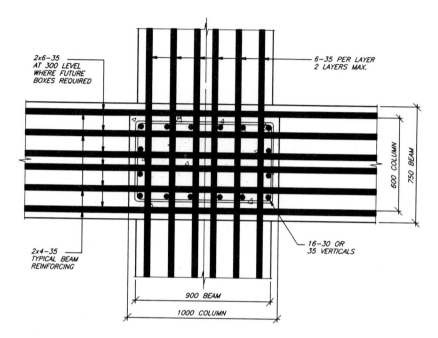

Fig. 4. Plan view of typical beam to column joint detail.

Coordination of reinforcement details with formwork system

As stated earlier, the coordination of reinforcement details with the steel formwork system was critical. The steel formwork impacted the placement of reinforcement in two key ways:

1. The steel formwork was only practical for the main frame beams and columns. This required that they be poured prior to the secondary beams. This resulted in a construction joint for the secondary beams at the faces of frame beams. In addition, the continuous bottom flexural steel required for frame action in the secondary beams could not be placed with the frame beam concrete pour. The solution was to install a sleeve through the beam at each location of flexural bars allowing these bars to be threaded through later.
2. The self-supporting steel formwork required the sides and soffits to be placed together. Side panels could not be omitted as is done with conventional formwork, to allow access for reinforcement placement. This required the beam reinforcement to be prefabricated in cages and craned into the steel formwork. For this reason, care was required in selecting the reinforcement detailing to conform with a logical division of prefabricated "cages". Of particular concern was the impact of column ties and beam stirrups on this sequence.

Figure 5 illustrates a typical column/beam joint which has the reinforcement arranged to accommodate the above described requirements.

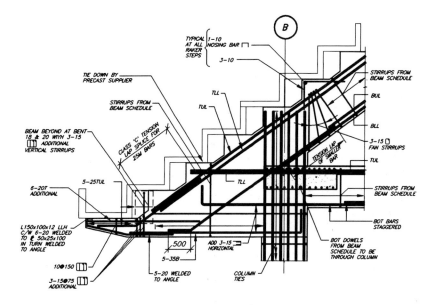

Fig. 5. Typical upper raker to floor beam detail.

STRUCTURAL DESIGN ISSUES

Strength versus stiffness design considerations

The stiffness, strength and stability of the Air Canada Centre Arena concrete superstructure is derived from a three dimensional moment frame. In addition, the lack of expansion joints in this relatively large building creates significant thermal and shrinkage stresses in the structure. The key to the member sizing and arrangement is to create a balance between stiffness and strength. Although a stiff structure will ensure manageable deflections it also has the effect of increasing thermal and seismic loads.

The process to determine appropriate member sizing starts with preliminary sizing for thermal, seismic and gravity loads. One should then perform a global thermal analysis and confirm the member sizing for seismic and gravity loads from full building computer models.

Thermal design

Loading

The thermal loading to be applied to structure is a combination of anticipated temperature differential and shrinkage strains converted to a ΔT using the thermal coefficient of reinforced concrete. To determine these strains we referred to the Canadian Prestressed Concrete Institute Metal Design Manual[2].

Shrinkage strains for the structure are calculated as the amount that occurs after the temporary separation strips in the structure are poured at 45 days. These are then adjusted for reduction factors associated with building geometry and material properties. Two sets of strains are required. Firstly, a short term strain occurring during construction to be combined with the unheated building temperature strain. Secondly, a long term strain to be combined with the final heated building temperature strain. The strain values are as follows:

Finally, the thermal and shrinkage loads are reduced to account for creep relaxation of the structure. This reduction factor for thermal differential is 1/1.5 and for shrinkage is 1/3. The two load cases to be considered are as follows:

$$\text{Short term} \quad \Delta T \; = \; 19.4°C$$
$$\text{Long term} \quad \Delta T \; = \; 15.7°C$$

The first case governs and was used in the preliminary and final analysis and design.

Preliminary Thermal Analysis and Design
For preliminary thermal design, firstly we determined the anticipated lateral deflection of the structure due to the 19.4°C. The building will contract from the centreline since the structure is approximately symmetrical for stiffness.

The structural elements of primary concern are the columns. Shear failure of columns due to internal strains has been well documented. Slab and beam forces are primarily axial and can be managed through proper reinforcement detailing. For this reason, we concentrated on determining column moments and shears. By selecting a column at the extreme end of the building and applying the anticipated deflection, forces were determined in the column. We assumed the deflection would occur primarily above grade, due to the restraint of the foundation walls up to grade. We then assumed the deflection would occur to a maximum over half the building height and remain constant above that to the roof. By selecting the first lift of column on grid 32 and grid W1, we anticipated a deflection of 5.5 mm over a 6-0 meter height. By using fixed end moments with that deflection, the moment diagram was determined. These moments are relatively high, but manageable and were a worst case.

Final Global Thermal Computer Analysis and Design
The final design was performed with a computer model using SAP90. The following assumptions were made:

1. Uncracked sections were used since creep effects were accommodated in the loading.
2. All floor slabs and walls were modeled as diagonal bracing elements. This was done to simplify the modeling and interpretation of results.

The results from the above were used for the final reinforcement details. Generally, the lower level floor slabs and beams were heavily reinforced due to the foundation wall restrains at these levels. At upper levels the reinforcing reduced to the point where flexure or minimum reinforcement governed.

The columns were designed for the moments and shears obtained. As an added precaution, the shear capacity related to the column nominal moment resistance was used as a minimum. This is consistent with seismic design philosophy.

Comparison of Preliminary and Final Thermal Design Results
The preliminary and final moments were compared. Our preliminary moments were somewhat larger than the final ones, but were in the same order of magnitude. This can be attributed to the rotational capacity of the frame joints which we did not consider in our preliminary analysis.

Seismic design
Preliminary design approach
The first stage of the seismic design process for an arena structure is to identify where the primary lateral building stiffness exists. The building section presented in Fig. 2 illustrates a typical "frame" comprised of columns, floor beams and seating rakers. The diagonal seating rakers act as braces making these radially arranged frames far stiffer than the column and floor beam frames oriented in the tangential direction. This understanding reveals that any lateral forces will distribute primarily to the radial frames.

The second concept which must be understood is that the period of the building is not entirely that of a moment frame. The diagonal seating rakers tend to behave similar to shearwalls. Some frame action still exists in the Air Canada Centre since one floor level, the suite level, does not contain diagonal rakers. The result is the period of the building is somewhere between that of a moment frame and a shearwall structure. For preliminary sizing, the shear associated with a shearwall system was used to distribute to the different levels and radial frames.

Final design approach
For the final seismic analysis and design, the preliminary member sizes were used to create a three dimensional computer model using SAP90 software. The following assumptions were made:

1. All sections were assumed to be uncracked for the determination of the fundamental building periods and the member forces. This will result in higher forces than cracked sections due to a shorter building period. We felt this prudent since not all members would crack during seismic loading.
2. Diaphragms were entered as shell elements to correctly model the ring-shaped floor plates which have some flexibility.
3. The building base was assumed to be the grade level. Single perimeter foundation walls restrained this level.
4. All building mass, including the roof, was entered as isolated lumped masses at each beam/column joint.
5. Two orthogonal directions of loading are applied and then results combined with SRSS.

The analysis process was done in three stages. Firstly, a modal analysis was performed to check our assumptions concerning the building period. Secondly, a spectral analysis was performed with an unscaled response spectrum from the National Building Code of

Canada[3]. From this computer run a dynamic base shear was achieved. The last step was to re-run the model with the input scaled to code base shear calculated with a period derived from the modal analysis. This yielded a set of member forces to be combined with gravity loads for design purposes.

A final review was done for deflection checks. All members were re-entered as cracked sections and the structure was analyzed dynamically and statically to ensure allowable drift ratios were not exceeded.

Preliminary vs. Final results

There are essentially two types of results to compare between the preliminary and final analysis. Firstly, there is the building periods. The dynamic modal analysis resulted in both orthogonal directions having fundamental periods of approximately 0.8 seconds with modal mass contributions of approximately 60%. This is consistent with our preliminary assumption of the structure's dynamic behaviour being between that of a shearwall structure and a moment frame structure. The periods associated with those assumptions were 0.56 seconds and 0.93 seconds respectively.

The second set of results to be used for comparison are the frame moments. Figures 6 and 7 illustrate these. Figure 6 is the moments resulting from a hand calculated distribution of shears to the various levels of the structure proportioned to the stiffness of one frame aligned in the direction of force. Fig. 7 is the dynamic analysis frame moments. Both of these include for all eccentricities.

The similarities between the member forces gave us a good degree of confidence in our computer model, and verified our understanding of the building behaviour.

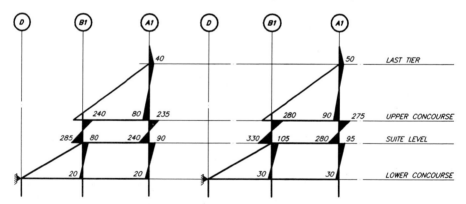

Fig. 6. Preliminary frame moments, Grid 24.

Fig. 7. Final frame moments, Grid 24.

REFERENCES

[1] Canadian Standards Association. S413 Parking Structures, 1994.
[2] Canadian Prestressed Concrete Institute. *Metric Design Manual*. 1987.
[3] National Building Code of Canada, 1995.

36 THE CENTRO ARENA, OBERHAUSEN, GERMANY

F. B. TIERNEY and G. JOHNSON
Bingham Cotterell, St Helens, UK

SUMMARY: The paper describes the structural engineering design solutions for the reinforced concrete frame and structural steel roof used for the construction of a 11,500 seat Arena in Oberhausen, Germany. Special emphasis is given to the use of precast concrete elements and the organisation and methods used by the construction industry in Germany.

Keywords: Acoustics, connections, engineering, frame design, German construction industry, design, groundworks, ice floor, ice hockey, multi-purpose stadium, precast concrete, roof structure, shear, stability.

INTRODUCTION

Centro is new centre for the City of Oberhausen near Dusseldorf in the State of North Rhine-Westphalia. Built by the Stadium Group, the complex includes a major retail centre, and associated facilities including the 11,500 seat Arena. The site of the Arena, which includes an ice hockey stadium, is a focal point for leisure in the development, which is linked to the old centre and surrounding districts by a light rail transport system. Buses and trams running every two minutes allow the public to embark and disembark in front of the Arena, promenade and shopping mall. Fifteen million people live within 1 hour travel time of the site.

The objective was to build a fully air conditioned multi-purpose Arena with an international 61.03 × 30.07 m ice rink with a high level of visual appearance and acoustic performance. It would be the first multi-purpose arena in Germany with a permanent ice floor separately covered for other events. This was achieved with credit to all involved and classical concerts, rock festivals, reviews such as Riverdance, galas featuring Placido Domingo, boxing and ice hockey have taken place since its opening in October 1996.

DESCRIPTION

The Arena is of 'end stage' configuration consisting of two decks with fixed seating for 10,060 spectators at an ice hockey event and has 22 suites. A plan is shown in Fig. 1.

The Arena was split into five zones for design control purposes, providing movement

Stadia, Arenas and Grandstands, edited by P.D. Thompson, J.J.A. Tolloczko and J.N. Clarke.
Published in 1998 by E & FN Spon, 11 New Fetter Lane, London EC4P 4EE, UK. ISBN: 0 419 24040 3

joints at no greater than 60 m centres. The building a has a 95 m-span roof constructed from 10 m-deep curved steel trusses spanning between in-situ concrete columns. These project up from a partial in-situ and precast concrete frame which forms the concourse and the raking terrace seating areas. Foundations are piled. The air handling system was a high level supply with nozzles and low level extract beneath the spectator seating. The construction of a builderswork plenum chamber below the concourse floor simplified the distribution of air back to the plant rooms. A general section is shown in Fig. 2.

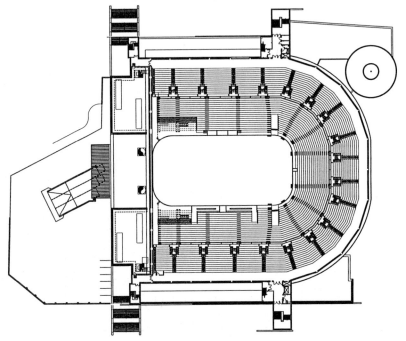

Fig. 1. Plan of Arena.

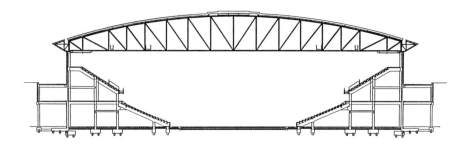

Fig. 2. Section.

General public access to the Arena is at concourse level, with stairs up to the upper terrace seating and direct access to the lower terrace seating. This is achieved by the creation of a man-made berm around the Arena held in place by a 7.6 m-high reinforced concrete retaining wall on piled foundations.

PROGRAMME

The concept design for the Arena was started in the summer of 1993 by the architects DLA Ellerbe Beckett in Wakefield UK, and Kansas USA. At the same time, demolition, decontamination and ground preparation for the site was started. The construction phase commenced with piling in September 1994, and the superstructure erection started in January 1995. The Arena was complete and handed over to the operators Ogden's in September 1996: a total of 40 months from inception to completion. Fig. 3 is a photograph of the Arena nearing completion.

Fig. 3. The Arena nearing completion.

DESIGN PROCEDURES

The procedure laid out by the German HOAI system is for the project to be advanced to Stage 3: 'Design development', before Stage 4: 'Approval of plans and calculations'. Due to programme requirements, the structural design was pushed ahead to Stage 4, in parallel with the architect's Stage 3 development. This required close co-ordination and joint design development with the architects in Kansas and the UK. Their requirements had to be implemented before the issue to the City authorities of development drawings. This fastrack approach may be common in the UK, but in Germany it is not normal and very precise methods and approvals for each stage of the process are followed. For structural engineering up to the start of the contract, these are as follows.

HOAI System
Stage 1: Basic scope
Stage 2: Preliminary design
Stage 3: Design development
Stage 4: Approval plans and calculations
Stage 5: Construction information
Stage 6: Preparation for contract

By law the calculations require to be approved by a proof engineer before construction can commence. Discussions with the proof engineers were held to agree principles and formats for submission before each stage of the design. The system is much more rigorous than in the UK and 'position plans' were required for each level to provide member design marks fully cross-referenced to the calculations. The approval and stamping of the calculations as Baufrei ('free for building') by the proof engineer is an important landmark for the project.

The production of Schaleplane or formwork drawings for Stage 5: 'Construction information', were produced to show every dimension so that the carpenters could create forms without reference to any other information. These drawings were approved by the architect and stamped Baufrei before the contractor could prepare the rebar drawings which in turn had to be stamped by the proof engineer before construction.

Due to the fastrack approach, Stage 6: 'Preparation for contract', was bypassed and tenders were sought on Stage 4 information. Although the contractor was appointed on the basis of Stage 4 information the design development and approvals were incomplete, and the contractor was thus able to introduce some alternative construction methods to assist the programme. At this stage the introduction of precast raking supports for the precast seating units was decided upon which led to some interesting and challenging connection details.

DESIGN DEVELOPMENT

The decision to use reinforced concrete, with a high proportion of precast components, was made on the grounds of speed and economy. Costing exercises were carried out by Bovis Tillyard, while comparative designs were done to ensure that the most economical methods were adopted. The choice of software packages chosen for the design was taken after consultation with our local partner, Nuhlen Statik.

Nematchek Allplus and Staad III were selected, both of which produce a range of design and drafting packages in accordance with DIN standards. The programs are very comprehensive and give shear and bending moment diagrams for a given section followed by areas of rebar. All calculation text and drawing text was in German, so the engineers and

Autocad technicians learned the technical aspects of German. A library of standard descriptions in both languages was created for use on drawings to simplify translations.

The work was predominately carried out in the UK but as the programme accelerated, and the contractor became more involved, the client's team and the contractor set up office in the same building near the site. This resulted in an excellent atmosphere with easy communication and exchange of views.

GROUNDWORKS

A very extensive ground exploration examination was undertaken. This showed that there was approximately 6 m of filled ground overlying the Emscher sands and marl, with a high water table from the presence of the nearby Rhine Heren canal and the Emscher river. It was decided that bored piles should be used for the foundations and that these should be founded in the sand layers.

The existing structures, mainly reinforced concrete and brickwork buildings, were decontaminated and demolished, and the debris then crushed and screened on site to produce hardcore. This material was incorporated in a ground model so that the building levels could be chosen to minimise the import and export of material. The material was used for the berm to externally raise the ground levels, and internally below the excavation of the sunken ice-floor.

There was continuous liaison during sitework with TUV, the environmental organisation appointed to check the decontamination process. Some small areas of undesirable material were found and removed from site but this is an expensive process in Germany. The bulk of the hardcore was used beneath the roads and buildings.

The fill was consolidated with vibrating rollers and a 1 m-thick platform of crushed material was laid over the footprint of all buildings prior to piling. A total of 1123 Franki piles was subsequently constructed for the Arena alone.

Every pile group was analysed with the German RZB program to give maximum compressive and tensile forces for each load condition taking into account lateral soil resistance. This was a much more sophisticated design than would be normal in the UK. Each pile was reinforced for the actual design load, tensile or compressive, rather than for a global allowance of overall capacity.

ROOF STRUCTURE

The roof structure consists of steel frames spanning approximately 95 m between concrete columns projecting from the main concrete frames. At support points the bottom boom of the truss bears on an elastomeric bearing pad bolted to the concrete column. Thus, the whole of the roof structure is 'floating' on top of the concrete framework.

Each frame was designed for vertical loading as a two-dimensional frame with spring supports, the spring stiffness values used in the analysis corresponding to the stiffness of elastomeric bearings. Members were sized in accordance with the German standard DIN 18800 using Staad III. Deflection was considered in detail and the trusses were pre-set on site to values counteracting the dead load deflection.

In terms of the overall stability of the roof structure, a three-dimensional analysis model was created using Staad III, into which notional horizontal loads and lateral wind loads were input. This model was also used to predict the likely deformation of the roof structure under a temperature variation of 25 degrees. The 3-D model was used for the design of all roof bracing for axial and self-weight forces, and to ascertain the axial forces in the top boom of all roof trusses from wind. Examination of the 2-D designs for the trusses was then carried out to check that the additional load from wind forces did not overstress any members. Deflection values and reaction forces at support points were then exported into spreadsheets for use in the design of the main concrete frames.

CONCOURSE CONSTRUCTION

The concourse is used as the primary diaphragm for the transfer of horizontal forces to shear walls, and has a 150 mm-thick in-situ reinforced concrete slab on top of precast 50 mm-thick panels. The secondary beams supporting the concourse were precast and continuous over the primaries where possible. The only soffit shuttering was therefore for the main in-situ beams.

CONCRETE SUPERSTRUCTURE

For the concrete superstructure a combination of moment frame action and both vertical and horizontal shear elements was used to achieve stability.

The design included simple and continuous beam design for elements taking vertical loading only. A two-dimensional frame analysis was prepared for the main stability frames with both vertical and horizontal loads incorporated, while a three-dimensional model was prepared incorporating all interacting shear elements transferring load down to foundations.

Frames cantilevered above the concourse level where the concourse floor slab was used as a horizontal diaphragm transmitting loads to vertical shear walls and shear cores formed at stair and lift towers.

MAIN STABILITY FRAMES

It was realised in the early stages of design that frame deflection and the practicalities of connection detailing would influence the extent to which precast construction could be used. It was with the maximum use of precast construction in mind that the frame analysis model was developed. A design requirement for fixity coupled with the construction preference for the intersection of precast elements at these locations, led to significant effort in connection detailing.

In accordance with the DIN 1045, for stability checks, second-order frame analysis was undertaken to account for secondary moments induced within the structure. Second-order effects tend to lead to larger columns, and frames become heavily reinforced. The computer system performs an initial analysis of the structure, with stiffness reduced to 30% of full capacity, under static loading, then amends the geometry in accordance with deflections calculated. This is iterated until convergence is reached when subsequent iterations have a negligible impact upon the structure. Forces and moments are therefore applied to the structure which are greater than those calculated by normal linear elastic analysis. Such secondary analyses can incorporate the effects of factors such as construction inaccuracy into the design process.

Apart from standard vertical dead and live loading, and horizontal wind load applied to external columns, there were some specific loads worth noting. Horizontal loads extracted from the three-dimensional analysis of the steel roof were applied to the top of the cantilevering columns and, in accordance with the German loading code of practice DIN 1055 for terraces, a notional horizontal loading equivalent to 5% of the total applied vertical load on the structure was applied horizontally.

Beam elements were designed within the two-dimensional frame model and columns were designed using the Allplus software package in order that additional forces could be included. Such forces included reactions from beams perpendicular to frames, and axial loads to perimeter columns from steel roof girders. These were applied eccentrically to the columns in accordance with the deflections of supports calculated in the roof analysis model.

PRECAST ELEMENTS AND CONSTRUCTION JOINTS

As mentioned previously the extensive use of precast elements within moment frames meant that special attention had to be paid to connections.

Fig. 4 shows the connection of the two raking upper terrace beams with the columns of the main frame. The only moment release at this point is at the end of the lower beam where a pin connection was modelled. Perpendicular to the raking members, and also intersecting at the joint is a longitudinal in-situ beam. The raking members were precast and the supporting column cast in-situ.

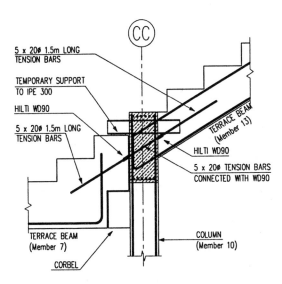

Fig. 4. Connection of upper terrace raking beams.

The construction sequence was devised to avoid craneage and temporary works, and an in-situ column was then cast to a level above the top of the corbel. Sockets cast into the column enabled corbels to be bolted on later. One beam was then placed onto the corbel to provide support, by means of a cast-in steel section. When this was in place the tensile reinforcement allowing the transfer of axial forces between the raking members was screwed into sockets previously cast into the end of the members. U-bar reinforcement left projecting was now within the column section providing full moment connectivity. The beam perpendicular to the raking beams was cast in-situ to complete construction of the connection joint.

Fig. 5 shows how the connection was achieved between the top of the lower terrace beam and the column of the main frame. Again, the in-situ column and in-situ beam parallel to grid line BB were cast first, and horizontal shear connectors were cast into the column. Beam 3 was precast with the top portion missing, as shown by the hatched area. The horizontal shear connector was bolted onto the component cast into the column and tie reinforcement screwed into the connectors. The exposed surface of the precast beam was then carefully prepared prior to the placing of the in-situ concrete to complete the structural element.

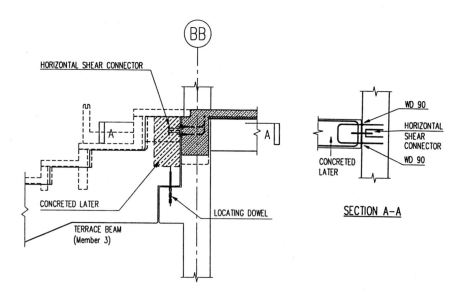

Fig. 5. Lower terrace beam connection with column.

STABILITY AND SHEAR ELEMENT DESIGN

Using the German computer program 'Aufteilung von Windlasten' the amount of total horizontal load taken by each vertical shear wall in proportion to stiffness was calculated.

Once the load in a shear wall was identified, pile groups and pile caps were designed accordingly. A combination of the horizontal resistance of the soil (which varied with depth) and raking piles was used to provide equal and opposite reactions to the horizontal forces applied.

The LEAP software package was used for the analysis of the large piled bases to stair and lift cores. The finite element model was composed of thick rectangular and triangular plate elements with out of plane bending and shear capacity. The nodes at corners of elements were released in only vertical translation so that unnecessary in-plane bending was not induced. Core walls were modelled as a series of continuous supports with vertical translational fixity only. Beam elements were used to model the position of these walls (i.e. linking the supports), with beam elements of infinite stiffness used to form rigid links. The location of piles were modelled as patch loads over an area equivalent to the pile cross-sectional area, with the total value of the patch load equivalent to the pile capacity. Linear elastic analysis of the model was carried out and the bending moment to be used for reinforcement design calculated based on the 'Wood and Armer' equations. Results were plotted in the form of contour lines representing the top and bottom reinforcement areas and bending moments in major and minor directions.

ICE FLOOR CONSTRUCTION

The construction of the ice floor is as shown in Fig. 6 and was designed to comply with DIN 18036 'Ice Sports Facilities Rules for Planning and Construction' and VDI 2075 'Technical Requirements of Ice Sports Facilities'. A ground-bearing slab was used and measures to limit frost penetration were determined by considering the one hundred year high groundwater level. If no measures were taken to limit frost damage then frost would penetrate in accordance with the following table.

Time (days)	10	20	100	150	200
Frost penetration (cm)	50	86	150	260	370

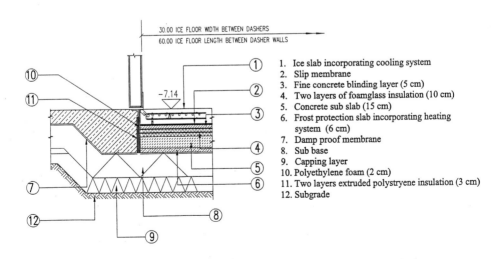

1. Ice slab incorporating cooling system
2. Slip membrane
3. Fine concrete blinding layer (5 cm)
4. Two layers of foamglass insulation (10 cm)
5. Concrete sub slab (15 cm)
6. Frost protection slab incorporating heating system (6 cm)
7. Damp proof membrane
8. Sub base
9. Capping layer
10. Polyethylene foam (2 cm)
11. Two layers extruded polystryene insulation (3 cm)
12. Subgrade

Fig. 6. Ice floor construction.

Ice lenses could form within the fine-grained sub-base material leading to differential frost heave, and could damage the ice floor. A heating system was therefore installed between the sub-base and the ice floor construction. The ammonia coolant refrigeration pipes were placed 5 cm below the ice level, and the coolant piping fed into one large header trench at the stage end of the ice floor. This provided only one point of egress from the ice floor to minimise the risks from corrosion and provide easy access for maintenance. The ice support slab was 20 cm thick and designed for HA loading with an equivalent uniform distributed load of 33.3 kN/m^2. This bears on a slip membrane consisting of two layers with a friction value not exceeding 0.25, allowing for movement due to temperature change and contraction without stress. The insulation layer thickness below is designed to prevent heat transfer to the ice slab. Fig. 6 is a typical section through the ice floor construction.

ROOF ACOUSTICS

The roof construction was designed and tested in accordance with BS 52 210 to provide an economic solution to achieving a rating of 49 dB. The overall construction was 425 mm deep made up as the following legend to be read from inside to outside.

- Roof finish was the KAL-Zip aluminium standing seam system; h = 65 mm
- Thermal insulation course of mineral wool; density p = 30 kg/m^3
- Distance section (steel sheet t = 1.0 mm) with mineral fibre inlay; h = 165 mm
- Layer of gypsum fibre boards, t = 12.5 mm; density p = 15 kg/m^3
- Layer of mineral fibre boards; density p = 50 kg/m^3 ; t = 140 mm
- Layer of gypsum fibre boards, t = 12.5 mm; density p = 15 kg/m^3
- Vapour barrier of PE foil, loosely placed with overlapping joints.
- Internal filler of mineral wool to troughs of under deck; density p = 40kg/m^3
- Trapezoidal steel deck with web perforations Type TA 100, with glued acoustic fleece on the inside Type Viledon C1986 SP (glass fibre fleece)
- Decoupling through elastic inlay of granular rubber strips, t = 6 mm; Shore Hardness A < 50.

With this roof construction and treatment of the interior a reverberation time of less than 1.8 seconds was achieved.

CONSTRUCTION

In Germany it is normal for the design and drawing work to be completed before the contractor starts work on site. It was not possible in this instance to provide all drawings due to the speed of the development, and considerable emphasis had to be given to programme during construction. The fact that the contractor and the design team worked in the same location allowed changes to suit construction to be discussed and implemented to meet the demands of the programme.

An impressive point was the quality and volume of detailed planning that was applied by the contractors Phillip Holtzman and Strabag before work started. Every member and detail was analysed, and its appropriateness confirmed, before starting on site. This ensured that site operations were not affected by the fastrack approach to the design and that construction methods and efficiency on site was not prejudiced.

As a result of the degree of planning and thought, there was no major structural problems and the combination of in-situ and precast elements was justified. Supervision was the responsibility of the contractor but Bingham Cotterell on behalf of the client retained a watching brief and checked the quality systems employed on site.

The working relationship was based on mutual respect and on maintaining the highest quality in engineering. Sound engineering principles provide a common link and the co-operation between engineers from the consultants, the proof engineers, and the contractors, proved this was the case. Personal contacts were essential to meet the very demanding challenge of building the Oberhausen Arena with what was truly an international project team.

Acknowledgements
Thanks is given to the Stadium Group for their kind permission to publish this paper.

37 MANAGING MAJOR PROJECTS: SOME LESSONS FROM STADIUM AUSTRALIA – AN OWNER'S PERSPECTIVE

A. PATCHING
Tower Hill Investment Managers Ltd, Sydney, Australia
C. CHAPMAN
Stadium Australia Management Ltd, Sydney, Australia

SUMMARY: The paper reviews the key aspects of project management, with particular reference to the approach taken in the teams working on Stadium Australia. The need to change the culture in organisations to understand risk and to involve stakeholders is emphasised, along with the need for effective communication.
Keywords: Business As Usual (BAU), cascading communication, contractual issues, corporate communication, corporate culture, HSTI system, project management, risk, risk management, stakeholders.

INTRODUCTION

Project management **is** different from everyday organisation 'business as usual' management (BAU), but yet depends so much on effective BAU structure, procedures, and most importantly, culture, for its potential impact to be maximised.

The purpose of this paper is to address the key issues that are relevant to the Stadium Australia project and which we feel are equally relevant to and should be considered in the management of any project – and, indeed, are imperative to remain focused upon in the successful delivery of major projects.

We will not dwell on the all-important phases of the project management process. We presume these are known and accepted and will focus on five issues which can 'make' any project if properly attended, and will 'break' any project if ignored or overlooked:

Culture in the sense of the impact of the collective values of the people involved with the various organisations engaged on the project and the manner in which people from organisations with different (and perhaps incongruent) values sets interact with each other.
Risk in its normal sense and definition.
Stakeholders in the sense of the people or organisations with a vested interest in the project.
Contractual issues in the context of the contractual system selected in light of the factors mentioned previously.
Communication in the triple contexts of:
    (a) personal communications between those involved with the project;
    (b) corporate communications between stakeholders and/or project team members; and
    (c) formal contractual communication.

Stadia, Arenas and Grandstands, edited by P.D. Thompson, J.J.A. Tolloczko and J.N. Clarke.
Published in 1998 by E & FN Spon, 11 New Fetter Lane, London EC4P 4EE, UK.   ISBN: 0 419 24040 3

THE PARADOX

We began this paper with the assumption that the basic phase of the project management process would be understood and carefully observed by all practitioners. The reality is that, in most cases of our experience, they are not.

More specifically, the first and most important phase – the preparation phase – is often treated lightly or virtually overlooked. Yet it is here that the foundation for project management success is laid. What hope is there for that which is built upon poor or insufficient foundations?

The fact is that sufficient preparation time is a key factor in controlling all of the issues we are about to address. But preparation takes time, and time is money. What many people fail to realise is that, in the project management environment, lack of preparation time will almost inevitably cost more money in the long term than will the appropriate enthusiasm (and associated resource commitment) in the preparation phase.

The problem arises from people separating preparation from project delivery. In the mindset of **urgency** to get the delivery process (i.e. post-preparation activity) started, perhaps to avoid interest charges, perhaps to meet a completion deadline, or perhaps simply to satisfy over-enthusiastic delivery personnel, the **importance** of having a well-developed project plan is overlooked.

Virtually every project we have reviewed is progressed on a project plan comprising budget, schedule, scope and resources component elements. This simply is too limited for the major projects of the 1990s. For these, a project plan also encompassing feasibility or business case studies, risk identification and management strategies, stakeholder relationships, management strategies, approvals procedures strategies and communication strategies, is fundamental to success.

Let us now turn our attention to the key issues identified as having a most significant impact on major projects. The appropriate approach to management of these issues is to gain a very clear understanding of their scope and relative impact upon a specific project by dedicating the appropriate time and resources to the task in the all-important, all-too-often undervalued, preparation phase.

CULTURE

In his book *The Futureproof Corporation* co-authored with America's Dr Denis Waitley, Alan Patching defines project management as no more than "the delivery of results and not excuses".

Whether or not a project management operative, regardless of whether or not he/she is engaged on a major or a smaller project, can deliver results and not excuses, will depend primarily on two factors:

- the corporate culture, and
- the person's attitude.

For successful project management operations, the corporate culture must allow people to deliver results and not excuses. Corporate culture is a topic on which numerous books have been written and no doubt many more will be. The specific needs of project management in terms of BAU managers' behaviour can be summarised as follows:

1. There must be an acceptance at all levels of the corporate hierarchy that each senior person has two roles. The first involves issuing performance instructions to project management teams. The second involves them, having issued these instructions, becoming subordinate to the project manager to the extent that they, the instructing BAU managers, are required by the project managers to have input that will facilitate the project team delivering the stipulated results.
2. There must be at least an acceptance by senior management that innovation and productive suggestion can emanate from all levels of the organisation and not just from senior management levels. Indeed, this is an expectation of senior management of highly performing companies throughout the world.
3. Senior management must be willing to fully delegate responsibility and accountability (in the "your job's on the line" sense of this term) to project managers, and then to get out of the project managers way and let them deliver results, but never to accept excuses from those project managers.
4. There must be an understanding that corporate culture does not change until people's values do. The logic is simple. A corporation is no more than a collection of individuals. Therefore, the values or culture of a corporation will always be the collective values of the people that fill its ranks. However, many of the 'career values' of these people are often set to meet the expectations of senior management.

In this respect, for corporate values to change, first, senior management might need to change. The fact is, this isn't always a change easily forthcoming in organisations introducing project management. This is where the attitude of the individual project manager becomes of fundamental importance. Project managers who are results-focussed must have the attitude that they will always deliver what is expected of them regardless of the corporate culture they work within.

The construction industry is indeed fortunate as it has operated in an environment fitting the culture specification defined above for several decades. However, the issue of culture looms dauntingly on major projects. These jobs usually entail many vested interests – some financial, others without experience **within** the construction culture. It is here that culture clash occurs and culture shock results. Only strong leadership which focuses single-mindedly on the contractual or documented project goals, while attempting to bring cultures and attitudes into line with these goals by partnering, persistence or whatever it takes to deliver results in this task, will eventually achieve the desired outcome.

RISK

Risk is perhaps the most important focus of effective project managers. Often, either too little attention is paid to it, or too much of the wrong type of attention. For major projects in particular, either of these scenarios can produce disastrous results. Our belief is that appropriate attention to risk identification and management would satisfactorily deal with most of the problem areas that arise from the other 'key issues' addressed in this paper.

All too often, risk 'management' for major projects takes the form of a sensitivity analysis as part of the business case or feasibility study.

Such a sensitivity study takes the financial and other assumptions of the feasibility study and looks at the impact on the project financials of the project when they differ from the assumptions. It is nothing more than a "what if" analysis, albeit a most important one. For example, "what if the interest rate is 0.75% higher for the first 6 months of the project and the project duration is extended by 3 months?"

Sensitivity analyses are a most important part of the feasibility study process. However, it is folly to think of them as an effective risk identification and management process. At the very least, effective risk management should divide risk into quantitative **and** qualitative components and then address such questions as:

- What is the risk factor?
- When is it likely to manifest?
- What will be the indications of it manifesting?
- What impact would it have on the project?
- How can we avoid it occurring?
- Who will be responsible for this monitoring and avoidance action?
- Who will this person report to and how often?
- If the avoidance strategy fails, how will we mitigate the impact?
- Who will be responsible for the mitigation action?
- Who would this person report to and how often?

To be effective, the answers to these questions should be documented for each risk factor. This should occur even for quantitative risk elements which may have been analysed by software especially designed for this purpose (e.g. @ Risk; Monte Carlo modelling software etc). To avoid reams of paper which are never addressed, we suggest the full risk analysis exercise be condensed to a simple matrix format. An example of such a matrix appears as Table 1 and is consistent with the format used in the Stadium Australia project.

Even where traditional risk analysis is conducted for major projects, there appears to be an emphasis on quantitative aspects of risk. Nobody with an understanding of the risk content of the estimates formulated for most projects (which estimates are almost always unfortunately regarded as 'guaranteed maximum prices' by unrealistic and inexperienced BAU managers) would ever under-value the importance of quantitative risk analysis. Notwithstanding this, one should never be persuaded to overlook the importance of qualitative risk analysis. Here we evaluate the potential impact of such factors as:

- specific personalities in senior key project positions,
- form of contract to be used, and
- previous experience of the personnel on both sides of a contract (or from various divisions of a corporation) in projects of a similar size and nature.

In summary, we see the identification and control of risk as a key issue in the management of major projects, yet it remains an element of most projects to which little more than an 'identify and hope' approach is applied.

Table 1. Risk management matrix.

RISK FACTOR	LIKELY TO MANIFEST	INDICATORS	IMPACT ASSESSMENT	AVOIDANCE STRATEGY	RESPONSIBLE PERSON	REPORTING TO	MITIGATION STRATEGY	RESPONSIBLE PERSON	REPORTING TO	COMMENTS
Quantitative										
Estimate accuracy (example only)	At receipt of tenders	Wide tender range Tender different from estimate	Probably no more than 10%-15% of estimate price	Call selective tenders Arrange estimate Check pre-tender board report re extras or scope change pre-tender, if necessary	BZ	AP	To be devised on positive indicator	TBA	TBA	-
Qualitative										
Contractors Construction Manager	Commencement of structural steel design	Reluctance to accept changes High prices on variations Doubtful and unrealistic schedule dates set	Only 5% on construction price but could impact substantially on business plan	Involve all stakeholders on possible design changes Involve Director of construction company Try to involve contractor in small owner-ship capacity	JK	CJC/AP	Resort and partnering issue escalation system Approach contactor at Board level	AP	Board	-

STAKEHOLDERS

Stakeholders are those having a vested interest in the project. They fall into two categories:

(a) those having a financial investment or interest in the project; and
(b) those others having no direct financial investment in the project.

In our experience, stakeholders in category (a) are easy to identify but often difficult to satisfy and/or manage. Those in category (b) are often easier to manage, but their importance will only be underestimated at the project manager's great folly. A feature of category (b) stakeholders is that, while they are usually easy to identify, they are often easy to overlook in the ongoing management of the project.

For example, we consider that every person who is likely to attend Stadium Australia for an event is really a stakeholder and we are considering their needs in details such as:

- designing seating position with uninterrupted sightlines
- focusing on sufficient toilet facilities
- concentrating on catering facilities to enhance the user perception beyond a sporting event to a complete 'entertainment' experience.

We see the following as essential elements of the fundamental project leadership tool of stakeholder relationship management:

- offering full project disclosure while maintaining full and formal project instruction control;
- affording open project involvement and input while retaining final decision-making capacity;
- seeking full consensus where possible while realising that, in the end, individuals (project managers) make decisions; teams and committees don't. Of course, where the project manager's decision and the team concerns are congruent, happiness and satisfaction prevail. However, the next time a happy and satisfied team of shareholders suffers the effects of a wrong decision because it forced an incorrect view on an inexperienced or weak project manager will not be the first.

Leadership is not always demonstrated by an ability to be liked and admired.

Regardless of whether decisions align with stakeholders' wishes or recommendations, providing full and frank communication to such stakeholders of all decisions at all project stages is critical.

A word of advice to the inexperienced:

- Project management is not a magical cure for project or corporate politics. The experienced project manager expects dirty tricks, backdoor tactics and downright obstructive behaviour that some stakeholders can display when decisions do not go in their favour; and then deals with this behaviour professionally and unemotionally.
- Project managers who take any type of criticism or reactive behaviour personally (in response to decisions made) show their inexperience. They are also advised to have regular blood pressure checks.

A cynical sage (clearly an experienced project manager) once defined the real stages of project management as:

1. Euphoria
2. Disillusionment
3. Panic
4. Search for the guilty
5. Punishment of the innocent
6. Reward for the non-participants.

Approaching every project with a view that this may not be far from the truth will assist in keeping project managers out of the coronary care wards.

CONTRACTUAL ISSUES

We believe that this is one of the most misunderstood areas of project management, not just for major construction projects but for corporate projects of all types and sizes. This point is probably best illustrated by example.

The co-author of this paper, Alan Patching, was once commissioned to conduct a partnering workshop for a number of parties about to execute a contract for the provision of professional services. The client group consisted of a number of government departments. As a matter of procedure, Patching always conducts a 'contract audit' in such circumstances.

Such an audit involves first seeking the opinion of the parties concerning the status of the contract about to be executed. On this occasion, all agreed that they were happy to proceed with execution of the documents.

Patching then invited a member of one of the client group departments (who was a lawyer but who had not been involved in preparation of the contract) to conduct an exercise *only with the client group* and involving only three component activities:

1. A check of the client group members who were not party to the preparation of the contract documents to ensure that those documents were a true reflection of the requirements of the user group's brief.
2. An explanation of each clause of the contract to the client group members to check if this explanation was congruent with their perceptions of the document; and
3. A repetition of (2) with all parties to the contract present (including the service provider).

This contract audit revealed that:

1. There were 21 instances (in a relatively short and simple contract document) where the users felt the contract as written was not providing what their brief to those preparing the document required.
2. There were 43 points of the contract as explained upon which the client group parties did not agree.

3. Fortunately, the service provider could accommodate most of the requested changes.

The contract was eventually executed. However, what chance would that contract have had of effectively controlling risk (a prime reason for having a contract in the first case) if it had been executed with the client parties themselves not agreeing on 43 important details and not even knowing that they disagreed!

We believe that a contract audit immediately prior to the execution of documents and involving professionals who are not directly involved in their preparation is a fundamental key to the use of various forms of contract as a risk distribution tool.

It might seem unusual to some to see 'contractual' listed among our key issues for major projects. We believe that there is a general misconception about forms of contract and what they provide, among even very experienced practitioners in the property and construction industries. The same situation probably exists within other industries as well.

On Stadium Australia, we have sought to avoid the problem of determining precisely how much risk lies with whom by simply allocating 100% of the risk associated with various functions to specific contracting parties. For example, 100% of the risk for design, construction, operation and maintenance of the Stadium has been passed from the Olympic Co-ordination Authority to the Trustee of the Stadium Australia Trust. In turn, 100% of the design and construction risk has been passed to the building contractor and 100% of the maintenance and operations responsibility has been passed to Stadium Australia Management (SAM) of which Christopher Chapman, co-author of this paper, is CEO. SAM has passed 100% of the everyday operations and maintenance responsibility on to an internationally experienced operator and a commercial partner in Stadium Australia, Ogden International Facilities Corporation.

For effective project managers, a key issue in managing major projects will always be either passing on 100% of risk to organisations with a proven performance record in dealing with such risk or, at the very least, taking a very practical and not at all dismissive attitude to fully understanding exactly where risk lies under the formal arrangements that govern specific projects.

COMMUNICATION

Effective communication really is the mortar binding the component blocks of outstanding project management. Let us address the three components of communication necessary for effective project management in the order they were introduced earlier.

Personal communication

An unfortunate reality of life is that most of us live our lives supported only by the communication skills copied from our parents and siblings and our school teachers. The skill level gained from this exposure simply will not, in most cases, be sufficient for optimal performance in the high-level project management arena.

We will not go into the principles of effective communication here. We will simply note that it is our belief that project management practitioners who do not understand and practice the skills and techniques of transactional analysis, neuro linguistic programming, various question techniques, objection handling skills and (fundamental to all of the above) listening techniques, simply cannot become the best they are capable of becoming in their chosen field.

Corporate communications

We believe in very simple corporate communication principles for effective project management. These include the following.

1. Communications of a non-contractual nature ('conversational communication') should be open, frequent, unrestricted and multi-directional.
2. Contractual communication *must* be along formal contractual lines and preferably only between specific, formally nominated representatives of the various contracting parties. We sometimes refer to this as 'instructional' rather than contractual communication.
3. The cascading communication system is ideal for the project management environment and should follow the HSTI principle:

   H = *Hierarchical.* This is communication from board level to project management practitioners. It usually emphasises deliverables in point form only. It should be delivered verbally and supported by a written summary. It should go by the most direct route (and not via departmental structures) to the project management system.
   S = *System.* Specifically, the upper levels of the project management system within the organisation. Here the board instruction is expanded to cover whatever are the standard elements of usual instructive communication with the project management system. The instruction is then sent from the system (typically a project director) to the team. Again, the instruction is given verbally supported by a written summary.
   T = *Team.* Specifically, the team leader or project manager. It is here the communication is worked with detailed instructions covering all aspects of budget, feasibility, program, risk analysis, resources etc. The instruction is eventually passed verbally to the individual members of the project team responsible for delivery of various parts of the instruction.
   I = *Individual.* Experienced project managers often do not issue detailed written instructions to their individual team members. Rather they often request the team member to revert to them with a written outline of what they will do in response to the specific briefing.

The HSTI cascading communication system is simple, and it is recommended for all projects – simply because it works.

Formal contractual communications

These we have covered sufficiently earlier herein.

SUMMARY

Regardless of the size of project where a leader is challenged to manage, adherence to basic principles will always be fundamental to success. Beyond these a focus on the key issues of development of a culture within which project management can flourish, diligent attention to the identification and management of risk, managing relationships with stakeholders, not falling into the trap of regarding a contract as a safety net and the development of improved personal and corporate communication skills will be the foundation of growth and increased project productivity.

38 ENVIRONMENTALLY SUSTAINABLE DEVELOPMENT OF SPORTS VENUES

J. PARRISH
Lobb Sports Architecture, London, UK

SUMMARY: It should be every stadium developer, designer, operator and user's aim to continually improve and protect the environment. Energy and materials conservation needs to be an integral part of the development, design, construction, use, demolition and disposal of every stadium, new or existing.

Keywords: Construction, design, design optimization, disposal, energy, environment, environmental impact, flexibility, functional requirements, location, management, materials, planning, sustainability.

INTRODUCTION

In October 1997, the 160 nations which signed the Climate Change Convention at the Earth Summit in 1992 started their final round of negotiations in Bonn to seek action on climate change beyond the year 2000.

Their final agreement was disappointing and a triumph for vested interests and laissez-faire attitudes. But even the most far-reaching suggestions on the table fell far short of the scientists' demands for action to avert a series of climatic disasters including the disappearance of perhaps one third of Bangladesh, the Maldives and the Nile Delta; the melting of the Arctic ice; the diversion of the Gulf Stream to pass to the south of the UK and Northern Europe and the destruction of many of the northern and equatorial forests.

Almost a quarter of a century earlier in 1975 I handed in my final year architectural student thesis at Bristol University. The subject was environmentally sustainable development. The recent oil supply crisis created by the oil-producing nations had reminded us that the world's resources were finite and sustainability was the "in" topic of the day. My thesis, like many others at the time, set out the arguments for materials and energy conservation and the use of renewable energy sources. Society hasn't stood still since then, but today's recommendations for environmentally sustainable development are broadly the same as they were in the mid 70s.

When the majority of scientists tell us there is an environmental problem we should be addressing we know they are probably right. When the politicians finally agree to take steps to deal with the problem we know it's a major one and we should have done something about it a long time ago.

But it is not just a problem for the scientists and the politicians. They cannot solve it on

Stadia, Arenas and Grandstands, edited by P.D. Thompson, J.J.A. Tolloczko and J.N. Clarke. Published in 1998 by E & FN Spon, 11 New Fetter Lane, London EC4P 4EE, UK.   ISBN: 0 419 24040 3

their own. It is a problem for all of us. We are all responsible for taking care of our environment. This is particularly true of the decision makers and professionals here today whose decisions have a major impact on energy and materials use.

Environmental sustainability
Environmentally sustainable development (ESD), is becoming an increasingly important consideration for all future buildings. Care and respect for the environment should be an integral part of the procurement, design and development process, and should cover not only the practical aspects, such as materials, energy and waste, but also the functional and visual impacts of buildings and hence their effect on the quality of life.

When we talk of environmental sustainability, the first thing most of us think of is energy and materials conservation. We think of using efficient heat generators, good insulation, photovoltaic collectors and the use of renewable softwoods rather than tropical hardwoods. All these are important but when looking at stadia there is a far more fundamental consideration. *"White elephants are not environmentally sustainable"*. For a stadium to be environmentally sustainable it must be successful. For a stadium to be successful it must be needed, be built in the right place, be well designed to avoid obsolescence of function, be flexible, be well run and maintained and be financially viable.

Environmental impact stages
The environmental impact of a stadium can be broken down into the four main stages of its life; determination of function and location, design and construction, use and finally disposal. The environmental impact of the decisions taken at each of these stages is greatest in the first stage and then decreases through each of the following stages until at the time of demolition and disposal most of the impact has already been determined by earlier decisions. This presentation concentrates on the first two stages.

Environmental impact responsibility
Responsibility for environmental impact can also be attributed on the same basis and in the same descending order. The first stage being controlled primarily by the developers, their teams of advisers and the authorities; the second by the developers, their designers and the authorities; the third by the operators and users and the fourth and final stage by the demolition, disposal and recycling team.

Long-term planning
Wembley Stadium is about to be redeveloped to create a new national stadium fit for the 21st Century. In the excitement of anticipating the new, we should not forget that it is also celebrating its 75th anniversary this year. The existing stadium may not meet today's more demanding requirements, but it has given good service for most of its life and is a tribute to those who conceived, designed and built it. For three-quarters of a century it has escaped the trap of obsolescence of function. Many other stadia have not been so successful and even some critically acclaimed projects built within the last decade are already being described as obsolete because they cannot easily be adapted to satisfy society's changing expectations.

Designing a stadium for a long and successful future is the goal. But it requires long-term planning. Decisions based only on short term considerations often lead to early obsolescence and poor overall value. Care for the environment, like the development of successful stadia, is a long term challenge and also requires long term solutions.

In many instances functional sustainability and environmental sustainability are synonymous. But while in conventional accounting terms a non environmentally sustainable development can appear viable, an unsuccessful stadium cannot be environmentally sustainable.

STAGE 1: DETERMINATION OF FUNCTION AND LOCATION

This first stage has most impact on the financial and environmental success of a venue.

The most basic requirement for a sustainable stadium is that it must satisfy a need. Without public demand no development can be successful and every stadium must be able to host a programme of attractions that will ensure regular use of its facilities at close to full capacity.

Dedicated use by a sports team is generally a pre-requisite and the more successful the team and the greater their support, the more sustainable the venue. Full use equates with sustainability and venues need to be used all day, every day to optimise use of resources.

Perhaps the most dramatic improvement to sustainability can be made by sharing a venue between two or more teams. In Huddersfield the football and rugby league clubs chose to share the McAlpine Stadium. As a result the stadium became twice as successful, twice as sustainable.

For sports with similar playing areas sharing is easy. Sports with different requirements pose more of a challenge but this can be met, as with many arenas and the new stadia under construction in Sydney and Melbourne, by making the seating and playing areas flexible. A small increase in capital and running costs produces a dramatic improvement in utilisation.

The areas of a stadium outside the viewing bowl are equally important and the provision of appropriate sports related facilities and attractions enable operators to create a dynamic and welcoming environment to attract users between major events and to ensure a steady stream of additional income.

The events may be the most fundamental requirement for any stadium but the second most important consideration is almost certainly location. Its relationship with the community and ease of access have a major impact on its success and its relationship to the transport infrastructure and parking provisions will have a major influence on the use of private cars and their associated environmental impact.

The ideal sustainable stadium would probably be at the centre of its catchment area to minimise travel distances, in the centre of a major conurbation so that it can utilise existing mass transport infrastructure and pedestrian links and in an area where it can be a focus for the community and a catalyst for associated developments.

Above all the stadium must be a long term success and for environmental sustainability the developer and all those advising the developer must be committed to these goals.

STAGE 2: DESIGN AND CONSTRUCTION

Many of the key principles of stadium design were understood by the Ancient Romans and were used to great effect in their magnificent stadia such as the Colosseum in Rome. Even today, the quality of their designs and the power and excitement of their spectacular events can be appreciated by even the most jaundiced tourist. That they fell into disuse was due to major structural changes in society rather than to fundamental failings in design.

The rise and fall of the Roman stadia hold an important lesson for today's developers and designers. Success requires well-designed venues offering good viewing and facilities and an exciting atmosphere. Lasting success may also require long-term flexibility. If your venue relies totally on football and football loses popularity it becomes redundant. If it is sufficiently adaptable and can stage numerous other events it can thrive in spite of society's changing sporting tastes.

Design optimisation

All designs benefit from optimisation whether in the calculations of sightlines, the selection of durable, cost effective roof coverings or in the provision of toilet facilities. Design optimisation should be a key feature of sustainable development. If your sightlines are poor, or your row widths too narrow, your stadium will become obsolete earlier than a venue designed to appropriate standards. If your cheap roof covering fails and has to be replaced after ten years its cost in use and long-term environmental sustainability is probably lower than a more expensive but more durable roof.

Fortunately today we have wonderfully powerful tools to help the process of design and optimisation. Computers are cheap and amazingly powerful. Projects such as Stadium Australia in Sydney and the Millennium Stadium in Cardiff were designed from day one using the most modern computer techniques to optimise their form and construction. From the earliest stage of design they have been developed, modelled, viewed and tested in three dimensions to demonstrate and prove their designs.

Even mundane tasks, such as the assessment of cut and fill and adjustment of pitch and floor levels, respond to optimisation, and the cost, material and energy savings can be dramatic.

Design for long life and full use

The overall goal in designing a stadium is to produce a building that will have a long and useful life. Compact, intimate, safe viewing bowls, comfortable seating, good sightlines and exciting and dramatic structures and architecture create attractive and successful facilities. Inadequate facilities quickly become obsolete. While it may be possible to enhance support facilities such as catering and toilets, it is rarely possible to rectify poor sightlines, obstructed views, cramped seating or ineffective roof coverage and protection.

Whilst some of us believe we can anticipate the likely changes in stadium design over the next five to ten years we probably need to revert to our crystal balls to anticipate the changes that will occur during the 30+ years of a venue's life. The best way to minimise the risks of early obsolescence is to incorporate flexibility into the design of pitch and seating tiers allowing a wider range of events, into the concourses allowing changes in catering and entertainment facilities and into in the general and specialised accommodation surrounding a stadium allowing the venue to adjust and develop as society's requirements change.

As with adjustable seating tiers, retractable roofs widen the range of possible events by providing protection from rain, snow, cold and strong winds and by providing controlled lighting conditions for spectaculars such as concerts.

Environmental design and energy conservation

Energy costs account for a substantial proportion of the operating costs of a modern stadium. As with all building types these can be reduced by careful and innovative design and efficient use and the form of a stadium should be appropriate for the local environmental conditions.

Appropriate environmental standards for temperature, humidity, air movement, air changes and illumination have a major influence on energy consumption and need to be determined from first principles for each venue and each area of a venue. Environmental performance needs should be a fundamental part of the be considered at the concept stage and building forms and services should be optimised to suit their environment.

Reduced energy use

There are numerous techniques for using energy more efficiently that should be considered for all construction projects. However, the first and generally most effective starting point is to examine and reduce energy usage, particularly for cooling, heating and lighting. To the careful selection of environmental criteria should be added other measures such as the use of passive and natural ventilation rather than powered systems, reductions in the use of electricity for lighting by providing transparent areas of roof, automatic controls, energy efficient fittings, lighting voids and reflectors. The options considered should include those generally applicable to energy conscious design as well as techniques specific to stadia.

Each of these measures needs to be assessed on the basis of their true long term cost in financial and energy terms to confirm the savings are genuine and to avoid the "bottle bank syndrome". This describes the well meaning but sometimes misguided actions of consumers going out of their way to recycle their waste glass, but using more energy driving to the local bottle bank than is saved by the recycling process.

Environmental impact of selected materials

The selection of materials that are environmentally friendly initially appears straightforward. Hazardous substances such as asbestos, solvent based paints or pvc which are known to be harmful at some stage of their manufacture, use or disposal should be avoided. Sustainable resources such as softwoods should be selected in preference to tropical hardwoods and low energy materials should be used in place of high energy materials such as aluminium.

But again, in making our selections, we must be careful to avoid the "bottle bank syndrome". A standing seam aluminium roof appears to have a very high material energy content. But it also lasts considerably longer than a felt roof and can be recycled. A cost in use exercise can compare the two over the anticipated lifespan of the building but what is the energy content of the labour used to install, maintain, replace and dispose or recycle each of the alternatives?

Re-use of materials is highly desirable and obvious examples are the collection, storage and use of rainwater for pitch irrigation at venues such as the McAlpine Stadium in Huddersfield and Stadium Australia in Sydney.

Power sources

After examining the possibilities for reducing the use of power we can turn our attention to efficient sources of power. Renewable energy such as sunlight, wind, hydroelectric and geothermal is environmentally the most attractive. Low grade energy uses such as space and water heating are the easiest to satisfy and numerous examples of effective systems are in use around the world. Generation of electricity from renewable sources is also practical, as demonstrated by numerous hydro-electric power plants and the increasing number of wind farms. Even direct generation using photo-voltaic cells is now a proven technology, albeit at very high cost.

Major benefits are also available from the more efficient use of non-renewable energy sources and techniques such as combined heat and power can be both effective and cost effective.

The nature of stadia and their intermittent use makes the use of renewable and efficient energy sources a major challenge and examples have been hard to find. But this is changing. Stadium Australia was subject to a stringent energy and materials audit during its design and incorporates techniques such as stack effect cooling and the co-generation of electricity and hot water from gas to reduce energy use and maximise efficiency. All stadia, both new and existing, should be subject to the same careful analysis.

STAGE 3: USE

Managers and operators

Although the main decisions regarding the function, location and form of a venue are taken in the first two stages, responsibility for ensuring its success passes to the managers and operators once construction is complete.

Their overriding responsibility is to make the venue a success by creating the events that will attract spectators and by ensuring all parts of the venue are well and efficiently run. To remain successful stadia need to evolve in response to competition and changes in society's expectations. Understanding and taking advantage of these changes over the life of the stadium are the managers' and operators' greatest challenges.

Efficient management should include proper use of the building controls to minimise energy and water use and thorough and efficient maintenance based on long-term cost-in-use principles.

Efficient management also covers minimising environmental impact through careful choice of environmentally friendly packaging materials and recycling of materials such as aluminium cans and energy used for heating, lighting, cooling and transport can also be reduced by careful planning and scheduling of events.

Spectators and users

Accurately quantifying the environmental impact of energy and materials utilised in the construction and operation of a major sporting venue is difficult enough, but when we turn our attention to the impact of decisions taken by the thousands or tens of thousands of spectators at an event analysis becomes virtually impossible.

There is, however, one key decision taken by each spectator or user that collectively is probably as significant in energy conservation and environmental impact terms as any other decision taken throughout the lifespan of the vunue. *"Do I go by car or use public transport?"*

The significance of the decisions taken in the first stage on the location of the venue becomes glaringly apparent once the energy used for long journeys in large numbers of private cars is calculated and the advantages of inner city sites served by good public transport networks become obvious.

Managers, operators, spectators and users

Although many of the key environmental impact decisions have been taken in the first and second stages managers, operators, spectators and users all have a major impact on the sustainability of sports venues.

STAGE 4: DISPOSAL

The final stage of demolition and disposal is largely determined by the materials and construction methods chosen at the design and construction stage. By their nature, stadia have sophisticated long span structures and require advanced demolition techniques. Their large size allows efficient recycling of materials such as structural steelwork and aluminium roofing but creates a problem for disposal of unwanted materials. The use of dangerous and damaging materials should be avoided at the design stage whenever possible.

CONCLUSION

Few of us can be in any doubt that there is an environmental crisis and that the greenhouse effect alone will create climatic disasters within the foreseeable future. All of us have a responsibility to tackle this problem, particularly those responsible for large scale use of materials and energy such as sports facility developers, operators, designers and users.

Environmentally responsible sports venues are few and far between, if they exist at all. This needs to change and quickly. The Sydney Olympic Stadium has made a start by demonstrating that environmental principles can be incorporated into stadia being built today. Many improvements can be made by addressing the points in the four stages I have outlined but governments, developers, owners, funders, design teams, contractors, managers and users all need to be committed to environmentally sustainable development to gain the maximum benefit.

It should be every designer's aim to continually improve and protect the environment, to look at water waste and recycling, plant waste, balanced use of energy systems subterranean aquifer systems, storage of rain water and its use for irrigation and more. Not least, design buildings that enhance the land or streetscape and can be appreciated as graceful landmarks in tune with the built environment. Architectural excellence must be the goal that all sports venues aim for, remembering that the safeguard of the environment was previous generations great folly; it must be this generations resounding success.

Author Index

Subject Index

This index is compiled from the keywords provided by the authors of the papers, edited and extended as appropriate. The numbers refer to the first page of the relevant papers.

Also available from E & FN Spon

Aluminium Design and Construction
J.B. Dwight

Construction Net
A. Bridges

Designer's Guide to the Dynamic Response of Structures
A.P. Jeary

Leisure and Recreation Management
G. Torkildsen

Managing Sport and Leisure Facilities
P. Sayers

Multilingual Dictionary of Architecture and Building Terms
Edited by C. Grech

Multi-purpose High-rise Towers and Tall Buildings
Edited by H.R. Viswaneth, J.J.A. Tolloczko and J.N. Clarke

Programme Management Demystified
G. Reiss

Spon's Architects and Builders Pricebook
Davis, Langdon and Everest

Spon's Ground Maintenance Contract Handbook
R.M. Chadwick

Sports Architecture
R. Sheard

Sports Turf
V.I. Stewart

Steel Structures: Practical design studies
T.J. MacGinley

For details of the these books and other Spon publications please contact: The Marketing Department, E & FN Spon, 11 New Fetter Lane, London EC4P 4EE; tel: 0171 842 2001